Infinitival vs Gerundial Complementation with
Afraid, *Accustomed*, and *Prone*

Juho Ruohonen · Juhani Rudanko

Infinitival vs Gerundial Complementation with *Afraid*, *Accustomed*, and *Prone*

Multivariate Corpus Studies

palgrave
macmillan

Juho Ruohonen
University of Helsinki
Helsinki, Finland

Juhani Rudanko
Tampere University
Tampere, Finland

ISBN 978-3-030-56757-6   ISBN 978-3-030-56758-3   (eBook)
https://doi.org/10.1007/978-3-030-56758-3

Cover illustration: © Harvey Loake

This Palgrave Macmillan imprint is published by the registered company Springer Nature Switzerland AG
The registered company address is: Gewerbestrasse 11, 6330 Cham, Switzerland

ACKNOWLEDGMENTS

This book has benefitted from informant judgments provided by John Calton, Joseph Flanagan, and Mark Shackleton. Particular thanks go to Laura Herrmann, who was generous in lending us her native ear on numerous occasions. We also wish to thank the anonymous reviewers and the people at Palgrave for their cooperation in what has been a smooth and straightforward review process. Finally, the authors acknowledge their indebtedness to Mark Davies, of Brigham Young University, for his work in compiling most of the large corpora used in this book and for making a range of corpora easily available to those engaged in the scholarly study of English. The responsibility for any shortcomings in the book naturally remains with the authors.

CONTENTS

LIST OF FIGURES

LIST OF TABLES

CHAPTER 1

Introduction

Abstract This book investigates infinitival and gerundial complements of selected adjectival heads in recent English, with essential use made of multivariate analysis. Authentic data from electronic corpora are used, and this chapter introduces the head-based and the pattern-based approaches to collecting data from corpora. Both infinitival and gerundial constructions are viewed as sentential, and this view is motivated in the chapter. The notion of control is introduced, because subject control is applicable to the analysis of the infinitival and gerundial patterns studied. The chapter also identifies the patterns to be investigated, and provides examples of them. The focus of the book is on heads that select both types of sentential complement—more specifically, multivariate analysis is applied to corpus data in order to gain information on the factors bearing on the incidence of each type of complement and to assess the importance of each factor. The Choice Principle is introduced in the chapter as an example of a factor to be studied in later chapters of the book.

Keywords Infinitival complements · Gerundial complements · Multivariate analysis and its application · Subject control

© The Author(s) 2021
J. Ruohonen and J. Rudanko, *Infinitival vs Gerundial Complementation with Afraid, Accustomed, and Prone,*
https://doi.org/10.1007/978-3-030-56758-3_1

"Complementation," in the words of Noonan (2007: 101), "is basically a matter of matching a particular complement type to a particular complement-taking predicate." In this sense complementation, even if the term was not used, was a prominent theme in traditional grammars of English, including Poutsma (1929 [1904]). In more recent work complementation has continued to be a central concern in generative studies, with Rosenbaum (1967) forming a basic framework for many later studies, in part because of the well-defined system of phrase structure rules offered in his book. Work in the generative framework was mostly based on intuitions and informant judgments, but the advent of large electronic corpora in the last two or three decades has given a new dimension to the field. The studies in this book take full advantage of the availability of large electronic corpora. In addition, the approach here draws on statistics in a more sophisticated way than is the case in most other current work on complementation. In particular, the use of multiple regression analysis promises new insights into the incidence of different types of sentential complements in recent English.

Two broad approaches can be identified in work on complementation that takes advantage of corpus evidence. The two points of departure follow naturally from the fact that complementation concerns a relation between a complement-taking predicate and a certain complement type. It is possible to proceed from the point of view of a particular complement-taking predicate or from the point of view of a particular complement type. A complement-taking predicate contains, or consists of, a lexical category—a verb, adjective, or noun—and the lexical constituent in question may be called a head, with the head selecting a complement type. The first point of view may then be termed the head-based approach, and it proceeds from identifying and examining the types of complements that a particular head selects. As for the second type of approach, it takes a particular complement type as its point of departure, and then examines the heads that select that particular complement type. A complement type is a phrasal category and as a phrasal category it may also be viewed as a pattern. The second point of view may then be termed the pattern-based approach.

In a pattern-based approach, a search string is chosen that does not specify the heads that select the construction, and the approach can yield information for instance on the verbs, adjectives and nouns that select a particular pattern. By contrast, a head-based approach recommends itself for instance when a particular head selects more than one type of complement such that the resulting head-complement constructions are fairly similar in meaning.

Most of the chapters of the present book are of the latter type, with only Chapter 3 diverging from this approach in that it takes up an analytical issue that arises in the study of adjectives co-occurring with infinitival complements of two types. In the other chapters the focus is on a head that selects two types of morphologically and morphosyntactically distinct sentential complements that are nevertheless so close to each other in meaning that they have often been treated under the same sense in major dictionaries. The objective is then to inquire into the different factors that affect the choice of complement on the basis of a selection of corpus data. A range of such factors are considered in the several chapters of this book, and advanced statistics is brought to bear on the salience of such factors. As far as the present authors are aware, this is the first book-length study of non-finite complements in English with such an orientation. The findings are based on the heads investigated, and may at first sight appear to be narrow in scope. However, the information gained here on factors affecting complement choice is useful beyond those heads in that it can be taken into account in subsequent work when other heads, quite possibly emerging from the application of the pattern-based approach, are examined in subsequent research. Ultimately, the head-based and pattern-based approaches can be combined and applied jointly in order to gain a full picture of the system of English predicate complementation.

The heads to be investigated are the three adjectives *accustomed*, *afraid*, and *prone*. In the case of *afraid*, the variation is between *to* infinitives and what are in this book called *of -ing* complements. For initial illustrations, consider the sentences in (1a–b), both from the Strathy Corpus of Canadian English.

(1a)   I was afraid to hang up.
(1b)   ... Quebecers are not afraid of going it alone.

In the case of *accustomed* and *prone*, initial illustrations of the variation to be considered are given in (2a–b) and in (3a–b), respectively. (2a–b) are from the Hansard Corpus and (3a–b) are from the Corpus of News on the Web (Davies 2013), henceforth referred to as the NOW Corpus.

(2a)   People in the Services are accustomed to disregard political prejudices. (Hansard 1952)

(2b)   We are accustomed to fertilising the land on which we grow crops ... (Hansard 1955)

(3a)   McPhee is more prone to stress the agony of composition... (US, 2017)

(3b)   Frankie seems prone to trying almost anything. (US, 2017)

An important basic assumption made in this volume is that all of the sentences in (1a–b), (2a–b), and (3a–b) involve a non-finite complement clause and that the complement clause, an infinitival clause in (1a), (2a) and (3a) and a gerundial clause in (1b), (2b), and (3b), has its own understood (or covert or implicit) subject. The idea that infinitival and gerundial constructions have covert subjects was made by major traditional grammarians, including Jespersen (1961 [1940]: 140), and it is generally made in current work. It is not shared by all linguists, but the present authors recognize that there is considerable evidence for understood subjects, in addition to an appeal to the tradition represented by Jespersen. One piece of independent evidence has to do with the generally accepted principle of binding theory that reflexives need to have a c-commanding antecedent (see for instance Radford 1997: 115). The well-formedness of a sentence such as *Perjuring himself would not bother John* then shows that the non-finite subject of the sentence has a covert subject, because otherwise the reflexive would lack a c-commanding antecedent. (The essence of the argument goes back to Postal's (1970) pioneering article, but today it is appropriate to present the argument in terms of binding theory; see Landau 2013: 115.)

The assumption of understood (covert) subjects, generally made in mainstream work today, also makes it possible to use a straightforward definition of the notion of control from Duffley (2014), which explicitly presupposes a framework with covert (or understood) subjects:

> Control has to do primarily with the question of what determines the identity of the unexpressed subject of non-finite verbal forms such as the infinitive or the gerund-participle in constructions such as *Joseph tried to find a quiet place* and *Peter enjoyed going fishing in his boat.* (Duffley 2014: 13; the term "gerund-participle," used by Duffley, corresponds to the traditional term "gerund" used in this volume)

The notion of control is an important syntactic property shared by all the six sentences in (1a–b), (2a–b) and (3a–b). Even more specifically, it is possible to say that each of the six sentences exhibits subject control, because in each of them the higher subject determines the reference of the lower (understood) subject. That the six sentences exhibit control, as opposed to movement, is clear because in each of them the matrix subject receives a theta role, or a semantic role, from the adjective of the higher sentence, which precludes a movement analysis of each construction. (A slight elaboration of this point, relating to certain constructions with *prone*, is offered in Chapter 6.) Since the lower clauses in (1a–b), (2a–b), and (3a–b) are control constructions, their understood subjects may be represented by the symbol "PRO," used for instance in Chomsky's work (Chomsky 1986: 119–132). "PRO" denotes a pronominal element without phonological realization.

Regarding the make-up of the book, complements of the adjective *afraid* are the theme of Chapters 2 and 4. The variation in this case is between *to* infinitive and *of* *-ing* complements of the adjective. Several factors potentially having a bearing on the variation in question are introduced in Chapter 2, with data from the Strathy Corpus of Canadian English and the British National Corpus, the BNC. In the discussion of explanatory factors potentially impacting complement selection, attention is paid for instance to the Choice Principle. This principle has been proposed in the very recent literature on the basis of univariate analysis in the case of certain control adjectives, including *afraid* (see Rudanko 2015; Chapters 3 and 4; Rickman and Rudanko 2018: Chapter 4), but in Chapter 2 the status of the Choice Principle is investigated for the first time with the help of multivariate analysis. The principle amounts to associating *to* infinitive complements with agentive lower subjects, on the one hand, and *-ing* complements with non-agentive lower subjects, on the other. The chapter is intended to provide information on an aspect of complementation in Canadian and British English, but, more importantly from a methodological point of view, the chapter aims to inquire into the explanatory variables impacting variation between two types of non-finite complements. Whether the Choice Principle can stand the test of multivariate analysis is one of the central research questions in the chapter. The chapter also investigates the predictive value of a number of additional variables that have been brought up in previous literature. Some of them prove too infrequent to enable a reliable multivariate analysis of their effects in corpora of Strathy's size category. Therefore, Chapter 4 avails itself of a five-billion-word section of the NOW Corpus, coupled with a purposeful sampling approach, in order to not only assess the validity of the Choice Principle in a different dataset, but to also gain information on the explanatory role of a number of low-frequency syntactic features that could not be reliably analyzed with the smaller corpora of Chapter 2.

Sandwiched between Chapter 2 and Chapter 4 is a chapter on what have been called degree complement constructions as in (4), from the NOW Corpus.

(4)   They sit silently, too afraid to utter a word. (US, 2016)

Sentences of the type of (4) have been noticed in the literature (for instance, see Baltin 2006: 267–269), and in such sentences the infinitival complement is selected by the degree modifier *too* (or by some other degree modifier), and not by the adjective (*afraid* in the case of (4)). The present authors duly noted the presence of degree complement constructions when they analyzed the data for Chapter 2, but they also noticed that not all substrings where a *to* infinitive follows an adjectival head preceded by *too* are necessarily degree complement constructions. The need to make provision for an adjectival head to select a complement even when there is a degree modifier present is clearly indicated by sentences of the type of (5). Sentence (5) is from the NOW Corpus.

(5)   Most libertarians are cowering frauds too afraid to upset anyone to take a stand on some of the most important cultural issues of our time. (US, 2011)

Sentence (5) displays nesting of *to* infinitives, with the inner one (the one closest to *afraid*) being a complement of *afraid* and the outer one (the one on the right edge of the sentence) being a degree complement dependent on the degree modifier *too*. As far as the present authors are aware, this type of nesting of *to* infinitives has so far been overlooked in the literature, which provides a rationale for including the chapter in the present book. The chapter is also included in the present book because it offers comments on how to separate the *to* infinitives that are complements of adjectival heads from *to* infinitives that are complements of degree modifiers in the chapters of the present book.

Chapter 5 turns to the variation of the non-finite complements of the adjective *accustomed* in the Hansard Corpus. In this case the variation is between *to* infinitives and what are termed *to -ing* complements in the present book, and the focus is on the period from 1945 to 1964 as the period of the greatest variation. The chapter begins with comments on the syntax of the two types of constructions, illustrated in (2a–b) above, and it is argued that there is a sharp syntactic difference between the patterns of (2a) and (2b), in that in the infinitival complement of (2a) the word *to* is under the Aux node and that in the gerundial pattern of (2b) the word *to* is a preposition. In spite of the sharp syntactic difference the adjective *accustomed* has selected both types of constructions in recent times, and the chapter applies regression analysis to shed light on the variation. While *afraid* can be characterized as a dispositional or modal adjective, *accustomed* is aspectual in meaning (Quirk et al. 1985: 1228). One of the main objectives of this chapter, then, is to ascertain whether and how the importance of the explanatory factors identified in Chapters 2 and 4 differs between the two semantic types of adjective.

Chapter 6 introduces a third semantic type of adjective. Similarly to *accustomed*, the sentential complementation of *prone* alternates between *to*-infinitival and *to -ing* complements. This adjective has two important semantic features which are shared by *accustomed* and *afraid*, respectively. Its aspectual element is in ascribing tendencies and recurrent patterns of behavior. Its modal element is in the attribution of potential or propensity for an entity to act or function in a certain way. Given the adjective's multifaceted semantics, different usage events of *prone* give salience to different aspects of its meaning. Determining how this micro-polysemy might bear on complement selection suggests itself as an interesting analytical challenge. Another unique feature of *prone* relative to the other two adjectives is its ability to occur with inanimate and abstract subjects. Along with assessing the generalizability of the results of previous chapters to another semantic type of adjective, a new question to which Chapter 6 seeks answers is whether and how the animacy of the adjective's predicand may influence its complement selection.

REFERENCES

Baltin, Mark. 2006. Extraposition. In *The Blackwell Companion to Syntax*, Vol. II, ed. Martin Everaert and Henk van Riemsdijk, 235–271. Malden, MA: Blackwell.

Chomsky, Noam. 1986. *Knowledge of Language: Nature, Origin, and Use*. New York: Praeger.

Duffley, Patrick. 2014. *Reclaiming Control as a Semantic and Pragmatic Phenomenon*. Amsterdam and Philadelphia: John Benjamins.

Jespersen, Otto. 1961 [1940]. *A Modern English Grammar on Historical Principles. Part V: Syntax (Vol. IV)*. London: Allen and Unwin.

Landau, Idan. 2013. *Control in Generative Grammar: A Research Companion*. Cambridge: Cambridge University Press.

Noonan, Michael. 2007. Complementation. In *Language Typology and Syntactic Description*. Vol. II. Complex Constructions, 2nd ed., ed. Timothy Shopen, 52–150. Cambridge: Cambridge University Press.

Postal, Paul. 1970. On Coreferential Complement Subject Deletion. *Linguistic Inquiry* 1 (4): 439–500.

Poutsma, Hendrik. 1904. *A Grammar of Late Modern English. Part 1: The Sentence*. 2nd edition 1929. Groningen: P. Noordhoff.

Quirk, Randolph, Sidney Greenbaum, Geoffrey Leech, and Jan Svartvik. 1985. *A Comprehensive Grammar of the English Language*. London: Longman.

Radford, Andrew. 1997. *Syntactic Theory and the Structure of English. A Minimalist Approach*. Cambridge: Cambridge University Press.

Rickman, Paul, and Juhani Rudanko. 2018. *Corpus-Based Studies on Non-finite Complements in Recent English*. London: Palgrave Macmillan.

Rosenbaum, Peter. 1967. *The Grammar of English Predicate Complement Constructions*. Cambridge, MA: The MIT Press.

Rudanko, Juhani. 2015. *Linking Form and Meaning: Studies on Selected Control Patterns in Recent English*. London: Palgrave Macmillan.

CHAPTER 2

Statistics and Complement Selection: A Case Study of *Afraid* Based on Canadian and British English

Abstract A number of variables have recently been shown to play a role in the non-finite complementation of complement-taking predicates in Present-Day English. This chapter opens with an overview of such variables, then proceeds to analyze the magnitude of their effects in two case studies focusing on non-finite complementation of the adjective *afraid*, first in the Strathy Corpus of Canadian English, then in a parallel British corpus that was constructed by selecting appropriate sections of the BNC to match the composition of the Strathy Corpus. The variables under scrutiny include the Choice Principle and the Extraction Principle, passivization of the complement clause, text type, and negation. Mixed-effects regression modeling is employed to study the role of such factors in the alternation between infinitival and gerundial complements of the adjectival head. The Choice Principle is shown to be the foremost explanatory variable of the alternation, whereas the strongly correlated syntactic parameter of voice appears to be of negligible predictive value. There are indications that Fiction may be the most favorable register to infinitival complementation, especially in Canadian English, although this result must be treated as provisional until replicated in datasets that allow adjustment for author idiolect. Negation of the higher clause is also identified as predictive, but only in the British data is its effect statistically significant—furthermore, that effect seems to be conditional on whether the Choice Principle applies. Odds ratios consonant

© The Author(s) 2021
J. Ruohonen and J. Rudanko, *Infinitival vs Gerundial Complementation with Afraid, Accustomed, and Prone,*
https://doi.org/10.1007/978-3-030-56758-3_2

with previous research are observed for the Extraction Principle, but the low frequency of the phenomenon prevents a statistically significant result from being obtained.

Keywords Syntax · Non-finite complementation · Variation · Multivariate analysis

2.1   Introduction

Consider the sentences in (1a–b), from the Strathy Corpus and the British National Corpus, respectively:

(1a)   …They are afraid to make an independent move. (NF, 1935)
(1b)   …he is afraid of going to the country. (HHX)

Both (1a) and (1b) feature a sentential complement selected by the adjective *afraid*. In (1a) the complement is a *to* infinitive and in (1b) it is what is here called an *of* -*ing* clause, with the -*ing* form being a gerund. In each construction the lower clause has its own understood or covert subject. The understood subject is the subject of the lower predicate, and as was noted in Chapter 1, there is also other evidence for postulating an understood subject in a non-finite clause. Another similarity between the types of sentences illustrated by (1a) and (1b) is that both involve control. This follows from the fact that the matrix adjective *afraid* assigns a theta role to the subject in each case. Given that the constructions involve control, the lower subject can be represented with the symbol PRO, which is an abstract pronominal element that is phonologically unrealized. As regards the type of control exhibited by (1a) and (1b), both involve subject control. The two sentences may be bracketed as in (1a′) and (1b′)

(1a′)   [[They]$_{NP}$ are [[afraid]$_{Adj}$ [[PRO]$_{NP}$ [to]$_{Aux}$ [make an independent move]$_{VP}$]$_{S2}$]$_{AdjP}$]$_{S1}$

(1b′)   [[he]$_{NP}$ is [[afraid]$_{Adj}$ [[of]$_{Prep}$ [[[PRO]$_{NP}$ [going to the country]$_{VP}$]$_{S2}$]$_{NP}$]$_{PP}$]$_{AdjP}$]$_{S1}$

One difference between the two sentences, represented in the bracketings, is that the term "nominal clause" is applicable to (1b), but not to (1a).

The two types of complement constructions are so close to each other in meaning that they have often been treated side by side in standard dictionaries, including the *Shorter Oxford English Dictionary* (1993), where they are both under the sense "frightened, alarmed, in a state of fear." The similarity in meaning constitutes an incentive to inquire into their meanings and ways of separating them. Underlying such an investigation is what has been termed Bolinger's Generalization. This is the claim that a "difference in syntactic form always spells a difference in meaning" (Bolinger 1968: 127). The generalization sums up a principle that plays an important role in stimulating work at the syntax-semantics interface in current linguistics.

The original idea for the present chapter arose from Ruohonen and Rudanko (2019). That study inquired into *to* infinitive and *of -ing* complements of the adjective *afraid* in the Strathy Corpus of Canadian English, with a focus on the factors impacting the variation. The variation in question was studied with the help of multivariate analysis. The present chapter first reanalyzes the Canadian dataset through an improved and more fully developed application of multivariate analysis. Then it proceeds to apply the updated methodology to a parallel corpus of British English. The results of the study can be expected to yield information on the overall incidence and conditioning factors of the two types of non-finite complement selected by *afraid* in fairly recent Canadian and British data.

Regarding the structure of the present chapter, Sect. 2.2 details the composition of the Strathy Corpus. The statistical analysis of the data is carried out in Sect. 2.3. The section begins with univariate analysis, which leads to multivariate analysis, with the latter given a more extended treatment. Section 2.4 then presents the parallel corpus of British English that was compiled ad-hoc to facilitate comparison of the results obtained in the two datasets. Section 2.5 then sums up the main results of both analyses.

To conclude the present section, it is appropriate to introduce at least some of the factors or variables whose impact on complement selection is studied in this chapter. One of them, the Choice Principle, was already mentioned, albeit informally, in Chapter 1. However, it is much less well established than the Extraction Principle in the literature. The principle has been defined as follows:

In the case of infinitival or gerundial complement options at a time of considerable variation between the two patterns, the infinitive tends to be associated with [+Choice] contexts and the gerund with [−Choice] contexts. (Rudanko 2017: 20)

The two types of contexts are then a function of the semantic role of the lower subject: when that subject is an Agent, the context is [+Choice] and when the subject is not an Agent, the context is [−Choice].

In other words, the Choice principle is predicated on a link between semantic roles and complement selection, with the Agent role at the heart of the principle. This circumstance naturally calls for discussion of the Agent role. Interest in semantic roles arose in the late 1960s, and the original work was carried out by Gruber (1967) and Fillmore (1968). The former approached the issue from the point of an agentive verb, defining such a verb as "one whose subject refers to an animate object which is thought of as the willful source or agent of the activity described in the sentence" (Gruber 1967: 943), and the latter, using the term "case role" for what later is generally referred to as a semantic role, spoke of the Agentive as the "case of the typically animate perceived instigator of the action identified by the verb" (Fillmore 1968: 24).

In later work it has come to be recognized that the Agent role is not assigned by a verb alone and that instead the larger predicate needs to be considered. This insight goes back to work by Marantz (1984: 25–26) and Chomsky (1986: 59–60). The latter pointed to sentences such as *He broke a window* and *He broke his arm*. Of these the former is much more likely to involve an Agent as its subject than the latter. Another point to emerge more clearly in later work is that while Cruse (1973), for instance, spoke of nouns being Agents, it is more appropriate to say that NPs, rather than nouns, are Agents. The reason is simply that subject arguments of sentences are always (at least) phrasal.

Another important contribution to work on the concept of an Agent is Lakoff (1977). He considered "prototypical agent-patient sentences," and gave a long list of properties or features that characterize them. This approach introduced a concept of the Agent defined in terms of features, and this is adopted in the present work. Some of the properties that Lakoff identified are of limited applicability, for instance, number 14 in Lakoff's list is "the agent is looking at the patient," which seems narrow in scope, but an approach based on features is suited to bring out characteristics of agentivity in a well-defined way. In particular, three of

Lakoff's features are given prominence in the present treatment. These are numbers 4, 5, and 6 in Lakoff's list. These are given in the following, with Lakoff's numbers, but with the personal pronoun *he* changed to a gender-neutral expression in 5 and 6.

4. the Agent's action is volitional.
5. the Agent is in control of what he or she does.
6. the Agent is primarily responsible for what happens (his or her action and the resulting change).

The three properties characterizing an Agent are thus volitionality, control and responsibility, and they are characteristic of a [+Choice] context. For instance, the understood subject of (1a) is an Agent in (1a) and the context is [+Choice] in sentence (1a). It conceptualizes an event—the event of making an independent move—as something that the referent of the understood subject does volitionally, is in control of and is responsible for. The lower subject of (1b) is similarly an Agent. On the other hand, consider sentence (2), from the BNC:

(2)   I was afraid of being suspected of the murder, I suppose (G15).

The understood subject of the predicate *being suspected of the murder* in (2) is not an Agent and the context is therefore [−Choice]. The event of being suspected of a murder is conceptualized in the sentence as something that is not volitional for the referent of the understood subject of the sentence, as something that is not under his or her control and as something that he or she is not responsible for.[1]

The three features may be supplemented with other considerations. For instance, predicates with Agents as their subjects are more natural in imperatives than predicates lacking that property. For instance, to hark back to sentence (1a), an imperative in the case of the lower predicate of (1a) is entirely natural *Make an independent move!* On the other hand, *Be suspected of a murder!* seems less likely. Another consideration—one that goes as far back as Gruber (1967, 1976)—has to do with compatibility with purpose clauses, including the type *in order to* Verb ..., where the understood subject of the purpose clause is coreferential with the higher subject. Predicates with Agents as their subjects are easily compatible with them. For instance, *They spend money in order to keep the economy moving* is entirely natural. It is harder to imagine them in a [−Choice] context, as in *John inherited the money in order to get rich. (The example is from Gruber 1976: 161.)

Language is not logic and there may sometimes be hesitation when deciding whether a particular subject is an Agent, but the considerations given here are generally sufficient to identify Agents, in order to apply the Choice Principle. The principle has not yet gained an established status in the literature, and one purpose of the present study is to determine if the application of multivariate analysis confirms or disconfirms the principle.

Another variable affecting complement choice is what has been termed the Extraction Principle. The principle goes back to pioneering work by Rohdenburg and Vosberg, and it has gained a status in the literature that is more established than that of the Choice Principle. Here is Vosberg's definition of the principle.

In the case of infinitival and gerundial complement options, the infinitive will tend to be favoured in environments where a complement of the subordinate clause is extracted (by topicalization, relativization, comparativization, or interrogation etc.) from its original position and crosses clause boundaries. (Vosberg 2003a: 308; see also Vosberg 2003b: 202)

Vosberg's (2003a) definition only refers to the extraction of complements, but in later work adjunct extractions have also been noted in connection with the Extraction Principle in Vosberg (2006: 69) and Rudanko (2006: 43). Two examples are given in (3a–b), with (3a) being from the Strathy Corpus and (3b) from the BNC.

(3a)   Your women do what our men are afraid to attempt. (1994, ACAD)

(3b)   … a financially hard pressed charity like HealthWatch, which has to step in where the professionals seem afraid to tread. (FT1)

(3a) illustrates the extraction of a complement and (3b) the extraction of an adjunct. In each of them the gap is at the end of the string. More recently, Rohdenburg has observed that the *to* infinitive—he uses the term "marked infinitive" for it—"enjoys a privileged status in extraction contexts" and that the *to* infinitive outranks "all kinds of gerunds" in such contexts (Rohdenburg 2016: 481). The Extraction Principle is well established by now. However, it has not been examined using multivariate analysis so far, with the exception of Ruohonen and Rudanko (2019). It is an obvious research question to investigate its impact further with that method of analysis.

A third variable to be investigated is the potential role played by the *horror aequi* principle. This principle has been formulated by Rohdenburg. He writes:

> Very briefly, the *horror aequi* principle involves the widespread (and presumably universal) tendency to avoid the use of formally (near-)identical and (near-)adjacent (non-coordinate) grammatical elements or structures. (Rohdenburg 2003: 236)

In the case of an adjective pattern, the *horror aequi* principle concerns any potential dependency between the form of the verb preceding the adjective and the form of the verb in the following complement clause.

Other potential factors to be investigated include the effect on complement selection of insertions and of the active versus passive distinction in complement clauses. Another variable concerns the potential influence of negations on complement choice. Regarding negation, the authors make a distinction between *no*-negation (Tottie 1991) and the more frequently occurring *not*-negation, in order to obtain information on the potential influence of each of them on complement selection.[2] A further possible factor impacting variation is text type.

2.2 COMPOSITION OF THE STRATHY CORPUS

The Strathy Corpus of Canadian English is a 50-million-word collection of Canadian English extending diachronically from 1921 to 2011. Online access to the corpus, as well as the user interface, are provided by Mark Davies at Brigham Young University. Table 2.1 details the diachronic and genre composition of Strathy. As shown, the data are heavily concentrated around the turn of the twenty-first century, with about five-sixths dating from 1990 or later. The three best-represented genres are Academic, Newspaper, and Magazines, which together make up over 75% of the corpus.

Spoken texts are the next-largest category, but they mostly represent context-governed types of speech, such as commission hearings and board meetings. Fiction and Non-Fiction make up about 8 and 5% of the data, respectively, with the latter category mostly comprising rather technical

Table 2.1 Diachronic and genre composition of the Strathy Corpus of Canadian English

	1920s–1940s	1950s–1970s	1980s	1990s	2000s	2010s	Total
Spoken	0	0	0	94,527	5,592,381	187,689	5,874,597
Fiction	1,739,983	329,263	506,611	860,022	452,736	12,766	3,901,381
Magazines	0	0	1,388,416	2,185,009	6,359,030	55,358	9,987,813
Newspaper	0	0	835,569	1,805,388	9,948,930	510,807	13,100,694
Nonfiction	761,739	172,617	735,567	822,731	2,728	0	2,495,382
Academic	125,134	193,961	1,996,289	2,523,454	9,575,865	230,650	14,645,353
Misc	0	0	49,437	0	26,620	0	76,057
Total	2,626,856	695,841	5,511,889	8,291,131	31,958,290	997,270	50,081,277

Source Adapted from www.english-corpora.org

prose, such as government reports and pamphlets. There is also a quantitatively insignificant Miscellaneous category. All things considered, it seems reasonable to regard Strathy as a corpus of relatively formal English. The lopsided diachronic distribution of the texts limits Strathy's value for diachronic inquiry, but it should be a fairly representative corpus of neutral-to-formal Canadian English of the late-twentieth and early twenty-first century.

2.3   COMPLEMENTATION
OF *AFRAID* IN THE STRATHY CORPUS

2.3.1   *Descriptive Statistics*

Given that the online interface did not support regular expression searches at the time the research was being conducted, we sought to improve recall by querying the corpus separately for ten different strings, then aggregated the relevant hits from the ten searches into one dataset. Table 2.2 lists the ten search strings used, along with the number of true and false positives yielded by each search.

The idea behind the multiple searches was to retrieve not only all canonical constructions, i.e. cases where nothing intervenes between the adjective and *to/of* nor between the latter element and the lower verb, but to also find as many tokens as possible that involved insertions. As expected, however, most of the relevant hits were of the basic type.

The discarded tokens included a handful of duplicates, instances where *of* was complemented by an indefinite pronoun such as *anything*, and

Table 2.2   Search strings used to retrieve the data

String	Relevant hits	Irrelevant hits
afraid to *	383	9
afraid of *ing	103	13
afraid -!|. to	0	0
afraid to -[v?i*] [v?i*]	2	2
afraid -!|. -!|. to	2	2
afraid to -[v?i*] -[v?i*]	4	4
afraid -!|. of *ing	4	4
afraid of -!|. *ing	1	1
afraid of -!|. -!|. *ing	2	2
afraid -!|. -!|. of *ing	0	0
Total	501	37

complements of the type *afraid of bad things happening*, in which the gerundial clause has an overt subject. In addition, there were eight tokens featuring what have been classified in the literature as indirect complements (Huddleston and Pullum 2002: 55) or degree complements (Baltin 2006: 267–269). One is shown in (4) below:

(4)   ... the kids are too afraid to say anything to anybody. (1992, NEWS)

In constructions like (4), the *to* infinitive is licensed by the degree adverb *too* rather than by the adjective that it modifies. Such tokens were therefore excluded from the quantitative analysis.[3]

Figure 2.1 shows the diachronic development of the normalized frequencies of the two non-finite complements of *afraid* in the Strathy Corpus. A declining trend is clearly apparent. One potential explanation is the encroachment of competing quasi-synonymous adjectives, such as *scared* and *terrified*, into the domain of *afraid* (Rickman and Rudanko 2018: 15–52). Another, not necessarily mutually exclusive, explanation may be that *afraid* is increasingly selecting finite complement clauses.[4]

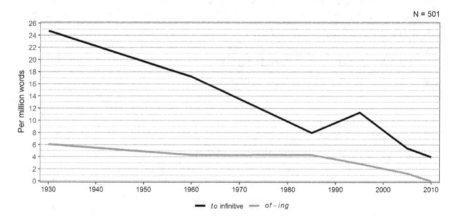

Fig. 2.1   Diachronic development of the normalized frequencies of the two non-finite complements in Strathy

2.3.2   The Limitations of Univariate Statistical Tests

There appears to be a strong correlation between [±Choice] and complement selection. Our dataset features a total of 396 [+Choice] contexts. Of them, a total of 369 involve a *to*-infinitival complement and only 27 an *of -ing* complement. In turn, the total number of [−Choice] contexts is 105, of which 24 feature a *to* infinitive and 81 a prepositional gerund. It is customary in such settings to employ Pearson's χ^2 test of independence in order to determine whether the association between the two variables is statistically significant. This test returns a χ^2 statistic of 242.73 and a *p*-value below .001, providing extremely strong evidence against the null hypothesis that complement selection is independent of [±Choice].[5] However, this result does not constitute proof that the relationship is causal. For instance, it has been shown by previous research (Rudanko 2015: 41–48) that Voice is also a major player in this syntactic alternation. Consider (5a–d) below, where (5a–b) feature active complements, while (5c–d) are passive:

(5a)   Linda is afraid to tell her grandmother about Mr. Flint's sexual advances... (2000, FIC)

(5b)   Taxi drivers were afraid to leave their parked cars, even briefly. (2003, NEWS)

(5c)   Growing up in relative poverty and afraid of being rejected, Lopez lied. (2003, NEWS)

(5d)   In short we're afraid of being attacked, torn to bits and eaten by bears. (2000, NEWS)

Passive complements seem to favor the gerundial variant. In the dataset at hand, the ratio is 18 *of -ing* complements to 3 *to* infinitives, while the active complements include only 90 gerundial variants as opposed to 390 infinitival ones. As expected, this association also proves highly statistically significant in the χ^2 test ($\chi^2 = 53.35$; $p < .001$).[6]

It will not have gone unnoticed, however, that while the complements in (5a–b) are active clauses, they also involve [+Choice] contexts. Conversely, the contexts in which the passive complements occur in (5c–d) are clearly [−Choice]. This is to be expected, since the prototypical function of the passive voice is to defocus the agent of a transitive clause, promoting the patient from object to subject.

What the simple univariate test of independence cannot answer is whether the observed selection pattern is due to [±Choice] or Voice. Both variables are heavily correlated with complement choice, but they are also heavily correlated with each other. Ascertaining the role of each variable independently of the other requires an approach which takes both of them into account simultaneously. This is where multivariate analysis offers a marked advantage.

2.3.3   Multivariate Analysis—Preliminaries

We used the *lme4* library (Bates et al. 2015) in R to fit mixed-effects logistic regression models (Hedeker and Gibbons 2006: 149–162) to the data. This is a variant of regression analysis, which is a family of multivariate methods designed to estimate the added value of each explanatory variable in predicting the outcome when the predictive value of every other explanatory variable in the model has already been taken into account (McElreath 2016: 123). In the present book, the outcome of interest is the occurrence of the *to* infinitive as the complement of the governing adjective, given that a non-finite complement occurs. Since probability is restricted to the [0,1] range, the presence of many explanatory variables combined with inevitable effect measurement uncertainty entails that the model will predict impossible probabilities unless the restricted-range problem is addressed somehow. Logistic regression achieves this by modeling probability indirectly via its *logit* aka log odds, i.e. the natural logarithm of the odds corresponding to the probability. Using the logit scale enables the effect coefficients to vary unboundedly while ensuring that the predicted probabilities, after the inverse logit transformation is applied, always stay between 0 and 1. Along with its other mathematically desirable properties (Agresti 2015: 142–143), the logit scale is advantageous in that exponentiating the effect coefficients yields odds ratios, which some find easy to interpret due to their use in sports betting. The odds ratio describes the estimated multiplicative change in the odds of the outcome corresponding to a unit increase in the value of the explanatory variable, when every other explanatory variable in the model is held fixed.

The term "mixed" refers to the inclusion of random effect(s) in the model. A random effect is a nominal-scale variable that is known or suspected to influence the outcome and has a very large number of discrete, unordered categories which typically continue to proliferate with additional sampling. The values of such unordered categories cannot easily be quantified for inclusion in the model as a single, numeric predictor, necessitating instead that an additional indicator variable be created to represent each additional category. This violates the regularity conditions of maximum-likelihood estimation (MLE), which require that the number of parameters remain fixed as N increases (hence the term "fixed effect" for typical explanatory variables). Violating this condition by modeling such a random effect as fixed results in parameter estimates that are biased away from zero, i.e. exaggerated (Demidenko 2013: 357–359). The random-effects approach assumes a common (usually normal) underlying distribution for the category effects of the problematic variable, shrinking their estimates towards the common mean in inverse proportion to their sample sizes. Then, the random effect constitutes only one additional parameter (the estimated standard deviation of the category effects), enabling the regularity conditions of MLE to be met and reducing bias.

In linguistics, the paradigm case of a random effect is idiolect. Unfortunately, this factor cannot be considered in the present book, since author information is not systematically included in the metadata of any of the corpora used. Another potential nuisance variable to consider for modeling as a random effect is the subordinate verb. Though it is by no means uncontroversial that verbs will vary in their baseline probability of the *to* infinitive vis-a-vis a prepositional gerund under government by an adjectival head, many linguists (Baayen 2008: 295–300; Levshina 2016: 252–253; Melnick and Wasow 2019: 14) do model the subordinate lexical item as a random effect. Moreover, there are examples in other languages of some verbs being much more prone to assume an alternate form for a given meaning than others (Brodsky 2005: 141). In this book, therefore, the identity of the complement verb is modeled as a random effect in order to control for potential lexical idiosyncrasies as well as to estimate their overall magnitude.

2.3.4 The Model-Selection Procedure

Throughout this book, we use a manual backward-elimination procedure to arrive at the set of explanatory variables included in the final model. The process starts from a model containing all explanatory variables of interest, proceeding in a stepwise fashion to remove variables that are not contributing predictive value and are not confounders. In assessing predictive value, we pay attention to both effect size and statistical significance, given that the two measures are not commensurate (see Sect. 2.3.7 below for details). Effect size is quantified by the absolute value of the estimated logit, i.e. the regression coefficient. For initial estimates of statistical significance, we use the Wald test due to its computational simplicity (Fox and Weisberg 2019: 322–323). This test compares the magnitude of a predictor's regression coefficient(s) to the estimated standard error(s), based on asymptotic normality. The test statistic, which we denote by W, has a large-sample chi-squared null distribution with degrees of freedom equal to the number of parameters introduced by the predictor (Agresti 2013: 10). More definitive assessment of a predictor's statistical significance is informed by the more reliable but more computationally expensive likelihood-ratio test, which is based on the difference between the model's goodness-of-fit with and without that predictor. This statistic is denoted by G, and it has an asymptotic chi-squared null distribution with degrees of freedom equal to the difference in number of parameters between the two models (Hosmer et al. 2013: 14–15). We cite G and the associated p-value before proceeding to drop a non-significant variable from the model and when reporting results for a final model.

To assess confounder status, we again follow Hosmer et al. (2013: 92, 109), whose criterion for confounding is whether a change of over 20% occurs in the coefficient(s) of any remaining variable(s) when the variable under consideration is dropped from the model. With quantitative variables (e.g. diachrony), checks for non-linear effects are always performed before the variable is definitively included in or excluded from the model. Once all superfluous variables have been culled, we always check for interactions among the remaining variables. An interaction occurs when the direction or magnitude of the effect of one variable is dependent on the value of another variable.

2.3.5   Model Selection

The dataset contained only six complements with insertion.[7] Three of them are presented in (6a–b) below:

(6a)   I don't like to tell you because I am afraid even to speak of it. (1928, FIC)

(6b)   All I could think of was what was behind my closet door. And how afraid I was to open it, just in case. (1988, FIC)

(6c)   I witnessed a brutal beating being inflicted by one schoolboy on another, so savage we were afraid not to stop and intervene. (1993, NEWS)

Two of the complements involving insertion occurred in [−Choice] contexts; they both selected an *of -ing* complement, whereas all four insertions occurring in a [+Choice] context selected a *to* infinitive. Thus, the small sample contained no information about the potential effect that insertions might have independently of [±Choice]. The same holds true for complement negation, i.e. the special case of insertion of which (6c) was the only example in our dataset. Since nothing can be inferred, let alone concluded, about the effect of insertions nor complement negation on the basis of such a small sample, we decided to set the insertion tokens aside, including the lone negated complement.

Only 13 tokens occurred in contexts where *horror aequi* might potentially apply. Seven of them involved a *to*-infinitival matrix verb, six a gerundial one. One example of each type is given in (7a–b):

(7a)   ...when you get an attachment, you have to be afraid to figure out where it came from... (2002, NEWS)

(7b)   ...she wanted to go into private psychotherapy practice after years of being afraid to quit her regular job... (2004, MAG)

In (7a), *afraid* selects a *to* infinitive despite being itself the complement of a *to*-infinitival VP. This runs counter to the *horror aequi* principle. Example (7b), on the other hand, does conform to *horror aequi*, given that *afraid* is immediately preceded by a prepositional gerund and subsequently selects a *to* infinitive as its own complement. However, complement choice in these two and all the remaining potential *horror aequi* contexts was perfectly predicted by the Choice Principle, which appeared to override the effect of the preceding context in our sample. There was thus little information about potentially independent *horror aequi* effects in the dataset, so we decided to exclude the *aequi* tokens from subsequent analyses.

A total of 15 tokens featured *no*-negation of the complement-taking adjective. While every one of them combined with *to* infinitive, all but one were also correctly predicted by the Choice Principle. Example (8) below was the only *no*-negated token whose complementation defied the Choice Principle:

(8)   They should never be afraid to find themselves alone because they have said what they believed to be true... (2000, SPOK)

Since nothing can be inferred from a single observation, we thought it best to leave the *no*-negated tokens out of the dataset. After the exclusion of both complement negation and *no*-negation of *afraid*, the only kind of negation that entered into the multivariate analysis was *not*-negation of the higher clause, an example of which is seen in (9):

(9)   Avorn is not afraid to confront shibboleths. (2005, ACAD)

After this final trimming of the dataset, we fit an initial model containing [± Choice], Voice, Extraction, *not*-negation of *afraid*, Diachrony, and Register as explanatory variables.[8] We first screened the variables for multicollinearity by calculating generalized variance inflation metrics for the fixed effects. The results are shown in Table 2.3. They key statistic is $GVIF^{1/(2*df)}$, which estimates the multiplicative increase in the uncertainty (i.e. standard error) around the parameter estimate(s) of each explanatory variable caused by the multiple correlation between that variable and all the other variables in the model (Fox and Weisberg, 2019: 429–434). This statistic was less than 1.4 for each predictor, suggesting that no severe multicollinearity was present.

A survey of the *p*-values and effect size metrics suggested that diachrony was the least consequential of the explanatory variables. Centered around the diachronic mean of 1987 and divided by ten to

Table 2.3 Generalized variance inflation metrics for the initial model

Predictor	GVIF	df	$GVIF^{1/(2*df)}$
Diachrony	1.94	1	1.39
Choice	1.48	1	1.21
Register	2.73	5	1.11
Voice	1.17	1	1.08
Negation	1.11	1	1.05
Extraction	1.09	1	1.05

represent estimated change by decade, the variable had a coefficient of virtually zero with a p-value of .86 ($W = -0.17$), indicating a total absence of any linear diachronic trend. As instructed by Hosmer et al. (2013: 95–96), we tested for non-linear effects by modeling time as a quadrichotomy, i.e. classified the data by diachronic quartiles and fit a model with a separate coefficient for each higher-order quartile. The results indicated some minor oscillation in complementation preferences throughout the observation period, but none of the quartiles had a coefficient greater than 0.2 in absolute value. Accordingly, the quadrichotomous variable yielded no improvement in fit over the simple linear term, which was itself statistically non-significant. We thus dropped diachrony from the model. This caused slight confounding in the coefficient of the Non-Fiction register, but the change occurred near zero (from -0.23 to -0.18), so we ignored it.

Voice was the next candidate for elimination, with a coefficient of only -0.23. Removing this variable had no adverse effect on the fit ($G = .07$; $df = 1$; $p = .79$) and caused no confounding, so we proceeded with the reduced model.

Register was first modeled as a senary taxonomy contrasting the default category of News with the five other registers present in the sample, i.e. Academic, Non-Fiction, Magazine, Fiction, and Spoken. Most of these category contrasts appeared superfluous, with near-zero coefficients suggesting negligible predictive value to show for the added model complexity. The main exception was Fiction, which had an estimated coefficient of 1.21 ($W = 2.29$; $p = .02$). We began simplifying the taxonomy by dispensing with superfluous category contrasts, first combining Magazine prose with the default category of News, then doing the same to the Non-Fiction and Spoken categories. These conflations had no adverse effect on the fit ($G = .24$; $df = 3$; $p = .97$), thus constituting a major improvement in model parsimony. However, trying to merge Academic prose with the reference category had a confounding effect on Extraction (whose coefficient increased by 20%), so we left it alone. Thus the final register taxonomy is a ternary one, with a reference category comprising News, Non-Fiction, Magazine prose, and Spoken material, contrasting with Academic prose on the one hand and with Fiction on the other.

The next candidate for removal was *not*-negation of the higher clause. While hardly negligible in magnitude, its coefficient of 0.6 fell short of statistical significance, with a p-value of .13 ($W = 1.49$). Closer examination revealed that out of the total 133 negated matrices, 120 (over 90%)

governed a [+Choice] complement. 112 of these 120 were *to* infinitives. While this is a heavy skew, it barely differs from the distribution observed for non-negated matrices with [+Choice] complements, which selected an infinitival complement in 227 out of 246 cases. More intriguing was the pattern observed with [−Choice] complements. Within this subset, 70 out of 87 unnegated matrices selected *of -ing*, while the corresponding fraction was only 7 out of 13 for negated ones. These numbers invite the hypothesis that *not*-negation of the higher clause may be a second-order infinitive-favoring factor whose main contribution occurs when [+Choice] does not apply. In fact, this is exactly what the fitted coefficients suggested when we added an interaction term for [+Choice] and matrix negation: 4.33 for [+Choice], 1.34 for matrix negation, and −1.03 for the interaction. However, just like the main effect of matrix negation, the interaction fell short of statistical significance as well ($G = 1.44$; $df = 1$; $p = .23$). The 13 negated matrices with [−Choice] complements were simply not enough data for reliable statistical inference. Since neither term was significant, we dropped matrix negation from the model altogether, leaving its closer investigation for later datasets that can quantitatively support such an endeavor. The simplification caused no confounding or significant worsening of fit ($G = 2.29$; $df = 1$; $p = .13$).

All the remaining variables either showed predictive potential or had been identified as confounders. As a final step, we checked for pairwise interactions between them. The only such interaction was found between [±Choice] and the Fiction register. In that register, the Choice Principle predicted all 78 complements correctly, resulting in an infinite positive coefficient for the interaction term. We felt such a register-conditional effect of [+Choice] to be highly implausible, i.e. likely a statistical fluke. We therefore stuck with the simpler model without interactions. The final model stands as follows:

Fixed effects:

1. [±Choice] (dichotomous)
2. Extraction (dichotomous)
3. Register (trichotomous—Other, Fiction, or Academic)

Random effects:

1. Subordinate verb (nominal-scale with 181 categories)

Before proceeding to model interpretation, we discuss the two criteria that this book uses to assess the predictive power of logistic regression models.

2.3.6   Model Performance: Classification and Discrimination

Perhaps the simplest and most intuitive way to quantify a binary regression model's predictive power is by calculating its "classification accuracy". The scare quotes are appropriate because the model itself does not "classify" the data, i.e. it does not produce binary predictions—it produces estimated probabilities, which are not dichotomous but continuous, albeit constrained to the $[0,1]$ interval. It is only through post-hoc dichotomization of these continuous values that we obtain binary predictions, which can then be compared with the observed outcomes in order to assess classification accuracy. Dichotomization necessitates that some probability cutpoint be chosen, so that data points with an estimated probability at or below the cutpoint are assigned a prediction of 0, and those with an estimated probability above the cutpoint receive a prediction of 1. Classification accuracy is then the proportion of these dichotomized predictions that matches the observed outcomes. The cutpoint is usually set at .5 (Hosmer et al. 2013: 170), and this book follows the same practice. For the model chosen in Sect. 2.3.5, this yields a classification accuracy of .898.

The forcible dichotomization of the scalar output of the model inevitably obscures important information about the model fit. Specifically, using the default cutpoint of .5 entails that classification accuracy cannot be lower than the sample proportion of the more frequent outcome.[9] Thus, in a dataset where one outcome category forms a 90% majority, classification accuracy will be at least .9 even if we use strings of random numbers as explanatory variables. One way to address this problem is by setting a cutpoint equal to the sample proportion of the "success" outcomes. Even then, however, it remains the case that the choice of cutpoint is arbitrary, and changing the cutpoint will yield a different classification accuracy. Using a higher cutpoint generally results in a higher **specificity**, i.e. probability that the model predicts a "failure" outcome for data points which have that outcome, but the tradeoff is a lower **sensitivity**, i.e. ability to correctly predict success outcomes for data

points with that outcome. Regardless of the cutpoint used, however, classification accuracy only provides a limited snapshot of the model's overall
performance, much as if we were trying to infer the overall performance
level of a soccer player from his or her performance in a single game,
rather than over the whole season.

A better and more complete measure of classification accuracy is
obtained by summarizing the model's overall ability to assign higher
probabilities to data points with success outcomes and lower probabilities
to data points with failure outcomes. This can be visualized by plotting
specificity against sensitivity over the entire range of possible probability
cutpoints, as in Fig. 2.2. This kind of graph is called a receiver operating characteristic (ROC) curve. While it is true in general that lowering
the cutpoint decreases specificity, it is equally true in general that for any
given cutpoint (and specificity), good models will have higher sensitivity

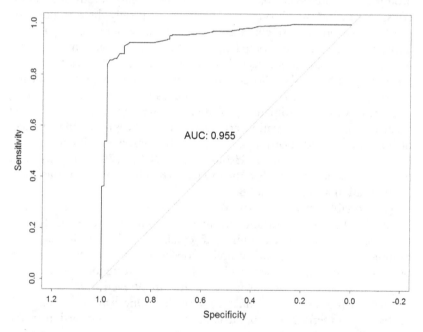

Fig. 2.2 Receiver operating characteristic curve of the model selected in
Sect. 2.3.5

than bad ones, because good models generally assign higher probabilities to data points with the success outcome than they do to data points with the failure outcome. By contrast, a useless model fails to discriminate successes from failures at all, so that as the cutpoint is lowered, increasing sensitivity is in perfect lockstep with decreasing specificity. This is represented in Fig. 2.2 by the straight diagonal line, which is how the ROC curve would look if the model had zero useful predictors and therefore zero ability to discriminate between successes and failures, even if its simple classification accuracy was good.

The area under the curve (AUC) summarizes the model's ability to discriminate between successes and failures. It ranges between a theoretical minimum of .5, as in a model whose discrimination power is no better than a coin flip, and a theoretical maximum of 1, which would occur if a model had perfect discrimination.

The area under the curve is equal to the so-called concordance index. For the full set of success-failure pairs of two data points that can be formed within the dataset, the concordance index is the proportion of concordant pairs, i.e. pairs where the data point with the success outcome has a higher estimated probability than the one with the failure outcome. For our model, the concordance index, i.e. area under the curve, equals .955. According to Hosmer et al., this is "outstanding" discrimination (2013: 177). However, good discrimination does not only depend on how well the model fits, but it is also a function of the magnitude of the regression coefficient(s) (2013: 174). A well-fitting model may only have modest discriminatory power if none of the explanatory variables has a particularly strong effect on the outcome. In the case at hand, however, our model discriminates very well because we have at least one very strong predictor. The next section discusses the estimated effects of the explanatory variables in more detail.

2.3.7   Model Interpretation

A key distinction to keep in mind when interpreting regression results is the difference between effect size and statistical significance. The former estimates the magnitude of an explanatory variable's impact on the response variable. It is expressed by the point estimate, and it is typically the main statistic of interest. Statistical significance, on the other hand, is inversely proportional to the p-value, and it measures our confidence in the observed effect being real, i.e. not due to random variation. Statistical significance is to a considerable degree a function of effective sample size. Effects of practically insignificant magnitude are often statistically significant when sufficient data back them up, and even dramatic effects are statistically non-significant when observed in samples of insufficient size.

Figure 2.3 shows point estimates and 95% confidence intervals for the fixed effects included in the final model.[10] The confidence intervals are inversely proportional to p-values, and an interval lying entirely on one side of zero implies statistical significance at the .05 level. The greater the distance between zero and the nearest endpoint of a point estimate's confidence interval, the more statistically significant the associated effect is.

All models presented in this book treat infinitival complements as the "success" outcome. The Intercept term in Fig. 2.3 represents the estimated log odds of "success" at baseline, i.e. when all the explanatory variables equal zero (if quantitative) or their reference category (if

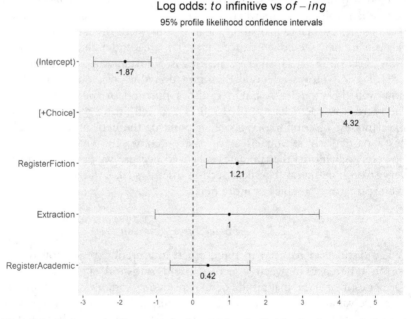

Fig. 2.3 Estimated effects on the log odds of *afraid* selecting a *to*-infinitival complement in Strathy

qualitative). This corresponds to a [−Choice] complement of *afraid* in the Informative register, with no extraction of constituents from the complement clause. The odds of such a token selecting a *to* infinitive are estimated at $e^{-1.87} = .15$, which equals about 13% on the probability scale. In analogous [+Choice] contexts, these odds are estimated to multiply by $e^{4.32} = 75$, yielding an estimated probability of $e^{-1.87+4.32}/(1 + e^{-1.87+4.32}) = 92\%$. This is the largest effect detected, and it is statistically significant at the highest, .001 level ($G = 138.77$).

In broad agreement with earlier work, contexts involving the extraction of a constituent from the complement are estimated to favor *to* infinitives at an odds ratio of $e^1 = 2.78$. However, the effect falls short of statistical significance ($G = .85$; $p = .36$). A closer look at the concordances reveals that the dataset contains only 14 extractions, of which 11 select a *to* infinitive. All 11 occur in [+Choice] contexts, however, so they provide hardly any information about the effects of Extraction independently of [±Choice] or vice versa. There are just three observations occurring in contexts where the two principles pull in opposite directions. They are shown in (10a–c) below:

(10a)   What she had been afraid of witnessing did not occur. (1930, FIC)

(10b)   ...dependency on others and physical and/or mental disabilities that we as individuals are afraid of having risk being translated as signifying a lack of worth... (2004, ACAD)

(10c)   "Where is the life you are so afraid to lose?" (1982, ACAD)

While the Choice Principle prevails over the Extraction Principle in (10a–b), resulting in an *of -ing* complement,[11] the opposite happens in (10c), where presumably due to the extraction of the direct object, an infinitival complement occurs in a [−Choice] context. It could be the case that the two principles both influence complement choice, but [±Choice] is the superordinate one that usually carries the day in cases of conflict. With only three observations, however, this hypothesis cannot be tested properly. The high *p*-value is a quantitative reflection of this incertitude. Extraction is simply too infrequent for its effect to be estimated reliably and independently of the Choice Principle in a dataset of this size.

The model estimates that with everything else equal, the odds of infinitival complements in the Fiction register are $e^{1.21} = 3.35$ times the odds in the reference category comprising Non-fiction, News, Magazine prose, and Spoken material. This seemed strange to us, and we suspected that the effect might be simply due to overrepresentation of some individual fiction author's infinitive-favoring idiosyncrasies in the Fiction sample. As far as could be ascertained, however, only two unexpected *to* infinitives (contravening the Choice Principle) were conclusively attributable to idiolect within this register. They are seen in (11a–b) below:

(11a) "He was afraid to get whipped." (FIC, 1966, Robert Kroetsch)

(11b) "He was afraid to be a fool. So he was a coward instead." (FIC, 1966, Robert Kroetsch)

However, given the absence of consistently marked author information in the metadata, we could not determine whether more individual-level clustering of this type was present in the data. While we recognize that there may be something about the Fiction register that favors the selection of infinitival complements, this hypothetical effect must be reproduced in other corpora before we can assume that it is not just an artifact of this specific sample.

The infinitive-favoring coefficient of Academic prose is small and statistically non-significant $(0.41; G = .58; p = .45)$. This parameter was kept in the model only to avoid confounding, and we are doubtful that any effect of substantive import exists.

As for the sole random effect, the estimated standard deviation of verb-specific logits is 0.74. This means that the average verb is estimated to favor one or the other variant at an odds ratio of $e^{0.74} = 2.1$ relative to the average verb. When contemplating this estimate, it should be remembered that unmeasured explanatory variables can contribute to its magnitude (Agresti 2015: 289–290). At any rate, this quantity can be used to estimate the intracluster correlation, i.e. proportion of the model's total explanatory power that is due to the random effect, i.e. lemma-specific idiosyncrasies. For the present model, it equals $0.72^2/(\pi^2/3 + 0.72^2) =$ 14%. The presence of the random effect improves goodness-of-fit by $G = 2.4$ $(p = .06)$, falling just short of statistical significance.[12]

The largest lemma-specific effect is observed with *get*, whose estimated odds ratio favors *of -ing* by a factor of $e^{0.91} = 2.5$ relative to the mean of all lemmas. Below are two examples of *get* contravening the Choice Principle.

(12a) He was afraid of getting this mad. (1985, FIC)
(12b) I don't like practicing the day of a game. I am afraid of getting hurt... (1983, NF)

The verb *fly* displays an even stronger gerund-favoring tendency in our dataset, with an odds ratio of $e^{1.08} = 2.94$. However, this is largely attributable to the fact that out of the total five tokens of this verb, three occur in the expression *afraid of flying*. Finally, *die* is a lemma whose semantics could be expected to strongly favor gerundial complements. This is not the case in our data, where six of its eleven complements are infinitival. One is seen in (13):

(13) I'm not afraid to die. Most transplant people understand. (1998, SPOK)

This verb has an estimated logit of 1.04, favoring infinitival complements by an odds ratio of $e^{1.04} = 2.82$ relative to the overall mean.

2.4 COMPLEMENTATION OF *AFRAID* IN A SAMPLE OF THE BNC

2.4.1 *The 50-Million Word British Counterpart to Strathy*

In order to compare and contrast the results of Sect. 2.3 with another major variety, we turned to the BNC to construct our own "British Strathy", i.e. a corpus of British English that was parallel to Strathy in both size and composition. Table 2.4 compares the composition of Strathy with its ad-hoc British counterpart. The compilation was a fairly straightforward task except for the matching of the two fiction subcorpora, where the BNC contains a significant sprinkling of historical fiction. In its emulation of speech from past eras, we suspected that this text type would be unrepresentative of the late-twentieth-century BrE of which the BNC was intended to be a sample. However, the historical fiction was mixed in with the other fictional prose, and the online user interface did not permit manual exclusion of individual texts from the searches, so we availed ourselves of a local copy of the BNC for finer control over the make-up of the subcorpus. To match Strathy's spoken subsection—which

Table 2.4 Register distribution of Strathy and the ad-hoc 50-million word "British Strathy" subcorpus of the BNC	Strathy	BNC Sample
Spoken	5,874,597	5,879,759
Fiction	3,901,381	3,966,733
Magazine	9,987,814	7,234,215
Newspaper	13,100,695	10,466,422
Non-fiction	2,495,382	7,829,478
Academic	14,645,353	14,653,047
Misc	76,057	65,511
Total	**50,081,279**	**50,095,165**

consists mainly of formal speech—we selected all BNC's Spoken categories except the informal *s_conv*, *s_demonstrn* and *s_sportslive*. Strathy's academic prose section was matched by selecting all BNC's academic registers except engineering. BNC's news section and magazine section (*W_pop_lore*) were both smaller than Strathy's, so they were included in their entirety. To make up the missing five million words, we made the non-academic/non-fiction prose section larger than Strathy's—after a glance at the subject matter in Strathy's non-fiction section, we selected the BNC's *W_nonac_arts* and *W_nonac_soc* sections in their entirety. To circumvent the problem of interspersed historical fiction in the BNC's fiction section, we used a spreadsheet program to randomize the order of the texts, then systematically hand-selected individual texts starting from the top and skipping historical fiction, until the cumulative word total matched that of Strathy's Fiction section.

2.4.2   Searching the Corpus

Our ad-hoc sample of the BNC was tagged with the Claws7 tagset (Garside and Smith 1997), with every token followed by an underscore and a part-of-speech tag (sometimes with several alternative tags). In order to retrieve non-finite complements of *afraid* from the corpus with maximum recall and reasonable precision, we used the following regular expression, in which the grayed-out segments are code comments:

```
(?i)(?<!\w)afraid_\S+\s(?!(,_,\s)?(and|or|but)_)((?![!.?<]_|that_)\S+\s)*?(
(?# to infinitive search begin)to_\S+\s((?![!.?<]_)\S+\s)*?\S+_v\w[^zg]\S*
(?# to infinitive search end)|(?# of -ing search begin)
```

of_\S+\s((?![!.?<]_)\S+\s)*?(?![kr]ing_|(any|some|no|every)thing_)\w+ing_\S+ (?# of -ing search end))\s

Without getting into much detail, this query retrieves instances of *afraid* that are not followed by a coordinated adjective or the finite complementizer *that*, but are followed by *to* or *of* which is followed by zero or more tokens of any kind except terminal punctuation, followed in turn by any verb form that might conceivably be an infinitive or a gerund.

The search returned 735 concordance lines. As usual, the many irrelevant hits included a large number of non-sentential complements as well as sixteen degree complements. A phenomenon not encountered in the Canadian data, however, was the presence of the chunk-like adverbial expression *I'm afraid to say*:

(14a) My first thought, I'm afraid to say, was 'What an ugly dog!' (ACM)

(14b) Speaker A: Does your husband help and things? Speaker B: Not very much I'm afraid to say, no. (GYY)

(14c) I enjoyed reading your material, but after consideration I'm afraid to say that I cannot make use of (unclear) short stories. (J9A)

These constructions, which numbered thirteen, were deemed complementationally invariant and were therefore excluded from the multivariate analysis.

After these exclusions, the total of relevant hits was 413, with 298 infinitival and 115 gerundial complements. This distribution is more even than that seen in the Strathy data—so much so that the overall difference between the varieties is statistically significant in the traditional (univariate) Pearson's χ^2 test ($\chi^2 = 5.09$; $p = .02$). However, we would not put much stock in this result, since it could well be due to an intervarietal difference in the distribution of important explanatory variables.

2.4.3 Model Selection

Unlike in Strathy, the effect of diachrony could not be estimated because it is not precisely or systematically marked in the metadata.[13] Regarding other potential predictors, there were only five potential *horror aequi* contexts—all of them with *afraid* complementing *to be*, and none with a preceding *of -ing* structure. Preliminary model fitting showed the former type to have a hefty but statistically non-significant gerund-favoring coefficient of 1.66 ($G = 1.27$; $p = .26$), suggesting that *aequi* effects may exist but cannot be reliably analyzed without more data. We chose to exclude the five *aequi* tokens from the dataset lest they confound the effects of other variables. There were likewise five instances of insertions preceding the complement. Three are shown in (15a–c) below:

(15a)   The United States er government is afraid as are other governments to admit er to their people that there is a higher form of life. (HV0)

(15b)   I suppose if he'd been alive, I'd be afraid now to meet him. (FS1)

(15c)   I was afraid not to look. (FSB)

The insertion in (15a) is a finite comparative clause, while the one in (15b) is a single adverb and the one in (15c) a negator. It is not clear whether these different types of insertion should all be grouped together, as we have done here, or whether they should be considered distinct phenomena from the standpoint of complementation choice. This is a question that cannot be answered without much more data. In any case, much like with *aequi*, the preliminary model reported a large but non-significant coefficient of 1.54 ($G = .67$; $p = .41$) for the insertion variable, suggesting that the phenomenon is worth further investigation. Meanwhile, we excluded the insertion tokens from the subsequent analyses in order to avoid confounding.

Thus, the initial model included [±Choice], Voice, Extraction, both types of matrix negation, and a senary taxonomy of Register into the categories of News (the default category), Speech, Magazine, Fiction, Non-Academic, and Academic. At this point we checked for multicollinearity among the explanatory variables, finding virtually none.

The two binary variables representing voice and *no*-negation, respectively, had each a negligible coefficient around -0.1, so these variables were promptly removed from the model after verifying that no confounding resulted. Then we looked into simplifying the Register categorization. As in the Strathy data, here too the most infinitive-favoring coefficient was observed for Fiction at 0.51, while the most gerund-favoring register was Non-Academic, with a coefficient of −1.13. The other registers all lay somewhere inbetween. Checking for confounding at each step, we ultimately subsumed Speech and Magazine under the default category and merged Academic with Non-Academic (whose coefficient had the same sign) to form a ternary taxonomy contrasting the default category with Fiction on the one hand and Learned Prose, i.e. a conflation of Academic and Non-Academic writing on the other. The binary variables representing the two non-default categories both had a coefficient of 0.72 with opposite signs. Although neither was statistically significant on its own, removing the whole variable caused a large increase in deviance ($G = 8.79$; $df = 2$; $p = .01$), so the trichotomy was kept in the model.

The next step was checking for interactions between the remaining predictors. Testing every possible pairwise interaction, our algorithm found a conditioning effect of [+Choice] on matrix negation, which appeared to favor infinitives dramatically in [−Choice] contexts, with a coefficient of 3, but only moderately in [+Choice] contexts, with a coefficient of 0.65. Directionally, this is the same phenomenon that was observed but fell short of statistical significance in the Canadian dataset (see Sect. 2.3.5). Despite its smaller overall size, the British dataset contained more matrix negation than the Canadian (153 vs 133 tokens). As a result, the interaction was significant in the British dataset ($G = 6.07$; $p = .01$), and its inclusion improved the model's concordance index from .928 to .942. The interaction was therefore kept in the final model.

2.4.4 Model Interpretation

Treating *to* infinitives as the "success" outcome, the final model is:

Fixed effects:

1. [±Choice] (dichotomous)
2. Extraction (dichotomous)
3. Matrix *not*-negation (dichotomous)
4. Register (trichotomous: Other, Fiction, or Learned Prose)

Interactions:

1. [+Choice] × Matrix *not*-negation (dichotomous)

Random effects:

1. Subordinate verb (nominal-scale with 150 categories)

The model has a classification accuracy of .864 and a concordance index of .942, discriminating the two outcomes very well. Figure 2.4 shows the estimated fixed effects and their 95% confidence intervals. The Intercept shows that in the "default" context with [−Choice], no extraction, no *not*-negation of *afraid*, and a register other than Fiction or Learned Prose, the estimated odds of an infinitive are $e^{-2.22} = .11$, translating to a probability of 10%. [+Choice] emerges again as the strongest predictor, increasing the aforementioned odds $e^{4.15} = 64$-fold relative to analogous [−Choice] contexts. *Not*-negation of the higher clause exerts a strong

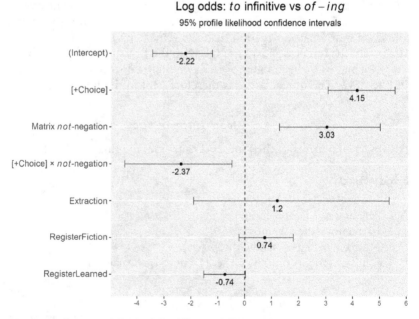

Fig. 2.4   Estimated logit-scale effects of the predictors included in the final model for the BNC data

effect conditional on [+Choice]. In [−Choice] contexts, the odds of negated matrices selecting a *to* infinitive are $e^{3.03} = 20$ times those of analogous non-negated ones. However, in [+Choice] contexts this odds ratio diminishes to $e^{3.03-2.37} = 1.92$. With Extraction we have a very similar situation as in the Strathy data: the lone extraction found in a [−Choice] context takes *of -ing*, while the seven others select the infinitive but are also agentive, so that there is little statistical information on the role of Extraction independently of [±Choice]. It would have been justifiable to drop Extraction from the model altogether, but we have retained it since its large infinitive-favoring odds ratio of $e^{1.2} = 3.33$ is consistent with earlier studies, even though it is not statistically significant ($G = .49$; $p = .48$).

As for the two registers being contrasted with the reference category of Speech/News/Magazine, the effects of both Fiction and Learned Prose fall short of statistical significance on their own, but the ternary variable is significant overall. The coefficient of Fiction has the same sign as in the Strathy data, and that of the Learned Prose category is likewise concordant with what theory would predict. We might indeed expect fictional narratives to use complement clauses to refer to specific actions, which have been associated with the infinitive, while learned prose may be expected to focus on generalities, which have been proposed as a salient semantic feature of the gerund (Taylor and Dirven 1991: 2, 17–19).

In contrast to the Canadian data, verb-specific effects seem highly significant in the BNC sample ($G = 8.2$; $p = .002$). The average verb is estimated to favor one or the other variant at an odds ratio of $e^{1.16} = 3.2$ compared to the population average. Expressed as an intracluster correlation, this means that 29% of the variance unexplained by the fixed effects is attributable to individual verbs. As in the Canadian data, the largest verb-specific effect is estimated for *get*, of whose seven complements five are gerunds. Three of them defy the Choice Principle quite conspicuously:

(16a) May we be open and honest and trustworthy so that other people are never afraid of getting close to us and need never be afraid of us hurting them. (ALH)

(16b) She discovered that she was afraid of getting close, afraid of being betrayed... (CA5)

(16c)   Not only are they afraid of getting in and out of the bath, but
        they also have problems in operating the taps and using the
        WC. (G2F)

The two *getting close* tokens were classified as [+Choice], but there is
arguably a connotation of helplessness in the expression, which may atten-
uate the infinitive-favoring effect of agentivity. By contrast, *getting in and
out of the bath* is unambiguously agentive. However, it is equally clear that
it refers to a general or recurrent phenomenon rather than to an individual event, which may explain why the gerundial variant was chosen.
Be that as it may, it seems clear that there is more variability in the British
data than in the Canadian—though the model has more predictors and
the dataset is slightly smaller, the amount of variability unexplained by the
fixed effects is greater. Whether this finding might be related to a tendency
for a colonial variety to show more internal homogeneity than its matrilect
(Romaine 1998: 23; Quirk et al. 1985: 17) is worth investigating in later
work.

2.5   CONCLUDING REMARKS

We have now investigated the role of six variables in the syntactic
alternation between infinitival and gerundial complements of *afraid* in
parallel corpora of Canadian and British English. Section 2.3.2 illustrated the significant univariate association of the complement selection
of *afraid* and the (lack of) agentivity of the covert subordinate subject.
Sections 2.3.5 through 2.4.4 then detailed multivariate, i.e. mixed-effects
logistic regression analyses of the impact of the Choice Principle and
a handful of other variables on this syntactic alternation in the Strathy
Corpus and the matching sections of the BNC.

Among the foremost results was the disentanglement of the Choice
Principle and passivization in the two multivariate analyses. Since the
passive voice defocuses the agent and moves the patient into the subject
position, subjects of passive clauses are prototypically non-agentive.
Therefore, it was previously unclear whether the observed association
between passives and gerundial complements of *afraid* was actually due
to Voice or [−Choice], i.e. whether it was due to syntactic or semantic
considerations. Our two analyses have provided a clear answer: at least in
the complementation of *afraid*, the Choice Principle is by far the more
salient factor, suggesting that semantics trumps syntax in this case.

Regarding the parameter of text type, in both datasets the Fiction register was estimated to favor infinitival complementation. The effect was more pronounced in the Strathy Corpus. These results should be viewed with caution. In particular, it is worth bearing in mind that neither corpus includes author information in the metadata of the texts, so there was no way to have our models account for the potential effects of idiolect. There thus remains a possibility, if remote, that the preference seen for *to*-infinitival complements in fiction may reduce to confounding by over-represented idiolects in that register. In any case, the impact of text type on complement selection will deserve further attention in later work.

The role that negation of the higher clause plays in its complement selection requires further investigation in larger samples. The parameter estimates were directionally similar for *not*-negation in both datasets, indicating that its infinitive-promoting effect seems to be at least partly conditional on the context being [−Choice]. However, only in the British dataset was the conditional effect statistically significant. This may conceivably be due to the larger presence of *not*-negation in that dataset. With regard to the less frequent *no*-negation, nothing could be inferred from the meager available data.

Our results on Extraction contexts were in broad agreement with the indications of previous literature. The odds ratio was decidedly favorable to *to* infinitives, but both datasets contained too few extractions to establish statistical significance.

It is much easier to point out a correlation than to prove cause and effect. When multiple factors are simultaneously at play in a syntactic alternation, analyzing each variable's conditional association with the outcome requires much more data than it does to obtain a statistically significant p-value for one variable's association with the outcome in a univariate test. Multicollinearity is a frequent phenomenon in observational data, such as corpora. Thus the more explanatory variables there are, the larger a dataset we need in order for there to be enough independent variation amongst them to facilitate reliable inference about their independent effects. An additional bottleneck to reliable statistical inference is often the low frequency of the rarer of the two outcomes (Hosmer et al. 2013: 408). This holds true for both of our datasets, in each of which the *of* -*ing* variant constitutes a clear minority. The problem is especially acute when analyzing relatively rare phenomena, such as extraction, insertions, complement negation, and the co-occurrence and differentiation of direct and indirect complements. Thankfully, massive corpora of very recent

English are nowadays at nearly everyone's reach (Davies 2013). The next two chapters make use of such a large corpus to inquire more thoroughly into the low-frequency syntactic phenomena on which the medium-sized corpora used in this chapter only enabled descriptive commentary.

NOTES

1. For some elaboration of the distinction between [+Choice] and [−Choice] contexts of infinitival and gerundial complements with *afraid*, see the comments in Rudanko (2014), Rudanko (2015: 46–47) and Rickman and Rudanko (2018: 28–29, 49, 65–67, 70–72). For another, very recent study of infinitival and gerundial complements of *afraid* and other fear adjectives, see Fisher and Duffley (2019). Their approach differs from that of Rudanko (2015), Rickman and Rudanko (2018) and of the present authors with respect to methodological choices such as analytical terminology, but there are also similarities and points of contact between the two approaches (see Fisher and Duffley 2019: 138). It will be a task for later work to compare and to evaluate the two approaches in more detail.

2. It has been proposed by an anonymous reviewer that an overarching parameter potentially influencing the complementation of *afraid* may be transitivity, as defined by Hopper and Thompson (1980). They conceptualize transitivity as a scalar property composed of ten features. Two of these features (i.e. agency and volitionality) overlap with the Choice Principle, while the rest are more orthogonal to it. The idea is presumably that higher transitivity would be expected to favor infinitival complements. Before this hypothesis could be tested, a careful discussion would need to take place on how to best operationalize the construct of transitivity for statistical analysis. That question deserves a separate treatment, and we leave it for future research.

3. We did not categorically treat all *to* infinitives occurring in structures of the type *too afraid to* (verb) as indirect i.e. degree complements. As an example, consider (i) below:

   (i)   Those who dare to take liberties will go highest in the dance…. Tradition is not enough…. Here there is this bright country but people are too afraid to try, too afraid to seem foolish. (1996, MAG).

We classified the infinitival complements in (i), and two others like them, as complements of *afraid* rather than degree complements. Premodification of *afraid* by *too* does not necessarily mean that the adjective must be followed by an overt, sentential degree complement. Rather, the degree complement may be omitted if its content is clear from the context.

Thus (i) could be rephrased with an overt degree complement, as in *too afraid of trying and seeming foolish to take liberties*. The other two cases in which *afraid* selected an infinitival complement despite being modified by *too* were similar.

4. This possibility was suggested to us by an anonymous reviewer.
5. We do not use a continuity correction with Pearson's χ^2 test, because they have a known tendency to make the test too conservative (Agresti 2018: 18).
6. Because of the very low number of prepositional gerunds in active complements, large-sample tests such as Pearson's χ^2 are not ideal for these data. A more appropriate procedure is Fisher's Exact Test (Agresti 2018: 46–49). In the case at hand, however, both tests return a p-value below .001.
7. We classified all constructions in which at least one word intervened between *afraid* and *to/of* as insertions.
8. To fit the models in this section, we used Nash and Varadhan's (2011) implementation of the Bobyqa optimization algorithm and 20 adaptive Gauss-Hermite quadrature points.
9. This ceases to be true in the fairly unusual event that the model has been fit without an Intercept term.
10. These are profile likelihood confidence intervals, which are more accurate than standard Wald intervals but much more CPU-intensive to calculate. See Hosmer et al. (2013: 15–20) for details.
11. The verb *witness* is compatible with both an agentive and an unagentive interpretation. In (11a), however, it is clearly unagentive. and could be glossed as 'see' or 'be the passive witness of'.
12. Given that the standard deviation of a random effect must be non-negative, the p-value of this likelihood-ratio test has been divided by two. See Agresti (2018: 278) for details.
13. Though the documentation at http://www.natcorp.ox.ac.uk/docs/ URG/BNCdes.html lists only 2% of the texts as having an unknown publication date, one of the present authors found over 20 percent of the texts to be missing a publication date in his close examination of the newspaper subcorpus for another research project.

REFERENCES

Agresti, Alan. 2013. *Categorical Data Analysis*, 3rd ed. Hoboken, NJ and Chichester: Wiley; John Wiley [distributor].

Agresti, Alan. 2015. *Foundations of Linear and Generalized Linear Models*. Hoboken, NJ: Wiley.

Agresti, Alan. 2018. *An Introduction to Categorical Data Analysis*, 3rd ed. Hoboken, NJ: Wiley.

Baayen, R. Harald. 2008. *Analyzing Linguistic Data: A Practical Introduction to Statistics Using R*. Cambridge: Cambridge University Press.

Baltin, Mark. 2006. Extraposition. In *The Blackwell Companion to Syntax*, vol. II, ed. Martin Everaert and Henk van Riemsdijk, 237–271. Malden, MA: Blackwell.

Bolinger, Dwight. 1968. Entailment and the Meaning of Structures. *Glossa* 2: 119–127.

Brodsky, David. 2005. *Spanish Verbs Made Simple(r)*. Austin: University of Texas Press.

Chomsky, Noam. 1986. *Knowledge of Language. Its Nature, Origin, and Use*. New York: Praeger.

Cruse, David A. 1973. Some Thoughts on Agentivity. *Journal of Linguistics* 9: 11–23.

Davies, Mark. 2013. *Corpus of News on the Web (NOW): 3+ Billion Words from 20 Countries, Updated Every Day*. Available online at https://corpus.byu.edu/now/.

Demidenko, Eugene. 2013. *Mixed Models: Theory and Applications with R*, 2nd ed. Hoboken: Wiley.

Fillmore, Charles. 1968. The Case for Case. In *Universals in Linguistic Theory*, ed. Emmon Bach and Robert Harms, 1–88. New York: Holt, Rinehart and Winston.

Fisher, Ryan, and Patrick Duffley. 2019. A Cognitive-Semantic Study of Verbal Complementation with Adjectives Expressing the Emotion of Fear in Canadian English. *Cognitive Semantics* 5: 121–136.

Fox, John, and Sanford Weisberg. 2019. *An R Companion to Applied Regression*, 3rd ed. Thousand Oaks, CA: Sage.

Garside, Roger, and Nicholas Smith. 1997. A Hybrid Grammatical Tagger: CLAWS4. In *Corpus Annotation: Linguistic Information from Computer Text Corpora*, ed. Roger Garside, Geoffrey Leech, and Tony McEnery, 102–121. London: Longman.

Gruber, Jeffrey. 1967. Look and See. *Language* 43 (4): 937–948.

Gruber, Jeffrey. 1976. *Lexical Structures in Syntax and Semantics*. Amsterdam: North-Holland Publishing Company.

Hedeker, Donald R., and Robert D. Gibbons. 2006. *Longitudinal Data Analysis*. Hoboken (NJ): Wiley-Interscience.

Hopper, Paul J., and Sandra Thompson. 1980. Transitivity in Grammar and Discourse. *Language* 56 (2): 251–299.

Hosmer, David W., Stanley Lemeshow, and Rodney X. Sturdivant. 2013. *Applied Logistic Regression*, 3rd ed. Hoboken, NJ: Wiley.

Huddleston, Rodney D., and Geoffrey K. Pullum. 2002. *The Cambridge Grammar of the English Language*. 3rd printing 2010. Cambridge: Cambridge University Press.

Lakoff, George. 1977. Linguistic Gestalts. In *Papers from the Thirteenth Regional Meeting of the Chicago Linguistic Society*, ed. A.B. Woodford, Samuel Fox, and Shulamith Philosoph, 236–287. Chicago: Chicago Linguistic Society.

Levshina, Natalia. 2016. When Variables Align: A Bayesian Multinomial Mixed-Effects Model of English Permissive Constructions. *Cognitive Linguistics* 27 (2): 235–268.

Marantz, Alec. 1984. *On the Nature of Grammatical Relations*. Cambridge, MA: The MIT Press.

McElreath, Richard. 2016. *Statistical Rethinking: A Bayesian Course with Examples in R and Stan*. Boca Raton: CRC Press/Taylor & Francis Group.

Melnick, Robin, and Thomas Wasow. 2019. Priming and Inhibition of Optional Infinitival to. Proceedings of the Ninth Workshop on Cognitive Modeling and Computational Linguistics.

Quirk, Randolph, Sidney Greenbaum, Geoffrey Leech, and Jan Svartvik. 1985. *A Comprehensive Grammar of the English Language*. London: Longman.

Rickman, Paul, and Juhani Rudanko. 2018. *Corpus-Based Studies on Non-Finite Complements in Recent English*. Houndmills, Basingstoke: Palgrave Macmillan.

Rohdenburg, Günter. 2003. Cognitive Complexity and *Horror Aequi* as Factors Determining the Use of Interrogative Clause Linkers in English. In *Determinants of Grammatical Variation in English*, ed. Günter Rohdenburg and Britta Mondorf, 205–249. Berlin: Mouton de Gruyter.

Rohdenburg, Günter. 2016. Testing Two Processing Principles with Respect to the Extraction of Elements out of Complement Clauses in English. *English Language and Linguistics* 20: 463–486.

Romaine, Suzanne. 1998. Introduction. In *The Cambridge History of the English Language*, vol. 4, ed. Suzanne Romaine, 1776–1997. Cambridge: Cambridge University Press.

Rudanko, Juhani. 2006. Watching English Grammar Change. *English Language and Linguistics* 10 (3): 31–48.

Rudanko, Juhani. 2014. A New Angle on Infinitival and *of -ing* Complements of *Afraid* with Evidence from the TIME Corpus. In *Corpus Interrogation and Grammatical Patterns*, ed. Kristin Davidse, Caroline Gentens, Lobke Ghesquière, and Lieven Vandelotte, 23–238. Amsterdam: John Benjamins.

Rudanko, Juhani. 2015. *Linking Form and Meaning: Studies on Selected Control Patterns in Recent English*. Basingstoke: Palgrave Macmillan.

Rudanko, Juhani. 2017. *Infinitives and Gerunds in Recent English*. London: Palgrave Macmillan Springer.

Ruohonen, Juho, and Juhani Rudanko. 2019. Comparing Explanatory Principles of Complement Selection Statistically: A Case Study Based on Canadian
English. *Studia Neophilologica* 91 (3): 296–313. https://doi.org/10.1080/
00393274.2019.1616215.

Shorter OED = *The New Shorter Oxford English Dictionary on Historical
Principles*. 1993. Edited by Lesley Brown. Oxford: Clarendon Press.

Taylor, John, and René Dirven. 1991. *Complementation*. Duisburg: L.A.U.D.
Linguistic Agency, University of Duisburg.

Tottie, Gunnel. 1991. Lexical Diffusion in Syntactic Change: Frequency as a
Determinant in the Development of Negation in English. In *Historical English
Syntax*, ed. Dieter Kastovsky, 439–467. Berlin: Mouton de Gruyter.

Vosberg, Uwe. 2003a. The Role of Extractions and Horror Aequi in the Evolution of -*ing* Complements in Modern English. In *Determinants of Grammatical Variation in English*, ed. Günter Rohdenburg and Brita Mondorf,
305–327. Berlin: Mouton de Gruyter.

Vosberg, Uwe. 2003b. Cognitive Complexity and the Establishment of -*ing*
Constructions with Retrospective Verbs in Modern English. In *Insights into
Late Modern English*, ed. Marina Dossena and Charles Jones, 197–220. Bern:
Peter Lang.

Vosberg, Uwe. 2006. *Die grosse Komplementverschiebung*. Tübingen: Narr.

SOFTWARE USED

Anthony, Lawrence. 2018. AntConc (Version 3.5.7) [Computer Software].
Tokyo, Japan: Waseda University. Available from http://www.laurenceanth
ony.net/software.

Bates, Douglas, Martin Maechler, Ben Bolker, and Steve Walker. 2015. Fitting
Linear Mixed-Effects Models Using lme4. *Journal of Statistical Software* 67
(1): 1–48. https://doi.org/10.18637/jss.v067.i01.

Harrell, Frank E., with contributions from Charles Dupont and many others.
2018. Hmisc: Harrell Miscellaneous. R package version 4.1-1. https://
CRAN.R-project.org/package=Hmisc.

Nash, John C., and Ravi Varadhan. 2011. Unifying Optimization Algorithms to
Aid Software System Users: Optimx for R. *Journal of Statistical Software* 43
(9): 1–14. http://www.jstatsoft.org/v43/i09/.

R Core Team. 2018. R: A Language and Environment for Statistical Computing.
R Foundation for Statistical Computing, Vienna, Austria. https://www.R-pro
ject.org/.

Robin, Xavier, Natacha Turck, Alexandre Hainard, Natalia Tiberti, Frédérique
Lisacek, Jean-Charles Sanchez, and Markus Müller. 2011. pROC: An Open-
Source Package for R and S+ to Analyze and Compare ROC Curves. *BMC
Bioinformatics* 12: 77. https://doi.org/10.1186/1471-2105-12-77.

Wickham, Hadley. 2016. *ggplot2: Analysis*. New York: Springer-Verlag.

CORPORA CONSULTED

Strathy Corpus of Canadian English. 2013. Created by the Strathy Language Unit at Queen's University. Available at: https://corpus.byu.edu/can/.

The British National Corpus, version 3 (BNC XML Edition). 2007. Distributed by Bodleian Libraries, University of Oxford, on behalf of the BNC Consortium. http://www.natcorp.ox.ac.uk/.

CHAPTER 3

Non-finite Complements of Fear Adjectives Nested Within *Too* Complement Constructions in Present-Day English

Abstract Previous literature on contemporary English syntax contains no article-length discussions of *to*-infinitival complement clauses nested within higher-level constructions with *to*-infinitival complements of their own. A case in point is the degree-complement construction headed by *too*. This construction expresses excess relative to a need, purpose, or desire encoded by a *to* infinitive, i.e. a degree complement, which may in turn be postmodified by a *to*-infinitival adjunct. Significantly, the degree complement may be omitted, and the *too* heading the construction may pre-modify a fear adjective (*afraid, scared* etc.) capable of governing its own *to* infinitive that is equally omissible. Consequently, considerable ambiguity may arise regarding the syntactic roles of *to* infinitives within, and at the periphery of, degree-complement constructions involving adjectives of fear modified by *too*. The present chapter draws on corpus data to illustrate these ambiguities, then goes on to propose semantic and syntactic criteria to aid in resolving them.

Keywords Syntax of fear adjectives · Non-finite complementation · Nested complements · Degree complements · Infinitival adjuncts

© The Author(s) 2021
J. Ruohonen and J. Rudanko, *Infinitival vs Gerundial Complementation with Afraid, Accustomed, and Prone,*
https://doi.org/10.1007/978-3-030-56758-3_3

3.1   INTRODUCTION

Consider (1a–b), from the *Corpus of News on the Web*:[1]

(1a)   Everyone quickly made it into classrooms, except for one little girl who was across the playground, too scared to move. (US, 2017, *Washington Post*)

(1b)   Everyone's too scared to speak up about it so I do in a jokey passive aggressive way. (IE, 2018, *The Irish Sun*)

The italicized sequences in (1a–b) denote excess in relation to some need, purpose, or desire (Huddleston and Pullum 2002: 1263). The *to* infinitive is a degree complement (Biber et al. 1999: 526), dependent on the constructional head *too*. The phenomenon falls into the broader category of what Baltin (2006: 267) has termed degree-complement constructions (DCCs). DCCs are Constructions in the strict Goldbergian sense (Goldberg 1995). This is evident in the fact that they can license the co-occurrence of a specific type of complement with lexical heads that do not normally admit that type:

(2a)   *I'm tired to proofread this wretched article.

(2b)   I'm too tired to proofread this wretched article.

The purpose of this chapter is to point to the co-occurrence or nesting of different types of *to* infinitives in the context of an adjective of fear, and specifically to inquire into the nature and identification of *to*-infinitival degree complements in relation to complements licensed by adjectival heads, mainly on the basis of data from a recent large corpus of English.[2] We became interested in the nesting of *to* infinitives in the context of a preceding degree word such as *too* when we noticed the well-formedness of sentences such as (3):

(3)   Most libertarians are cowering frauds too afraid to upset anyone to take a stand on some of the most important cultural issues of our time. (US, 2011, *Human Events*)

The syntactic relations obtaining in (3) can be represented in their relevant aspects by a bracketed string such as (3'). The symbol PRO denotes a covert subject that is co-referential with the matrix subject and lacks a surface form.

(3′) [[Most libertarians]$_{NP}$ are [cowering frauds [[too]$_{Deg}$ [[afraid]$_{Adj}$ [[PRO]$_{NP}$ [to]$_{Aux}$ [upset anyone]$_{VP}$]$_{S3}$]$_{AdjP}$ [[PRO]$_{NP}$ [to]$_{Aux}$ [take a stand on some of the most important cultural issues of our time]$_{VP}$]$_{S2}$]$_{DCC}$]$_{NP}$]$_{S1}$

The non-finite sentential complement that is directly governed by the adjective, exemplified in (3) by *to upset anyone*, is what will henceforth be called a fear complement.

Degree-complement constructions have been noted, sometimes under different labels, in major grammars of English, including Quirk et al. (1985: 1140–1142) and Huddleston and Pullum (2002: 1262–1264), but the kind of nesting of *to* infinitives observed in (3) has so far gone unnoticed. Given the absence of previous investigations, ours is intended as a first approximation to outlining the myriad complexities involved and offering some preliminary diagnostics to help bring at least a modicum of order into the chaos.

On the basis of informant judgments and Huddleston and Pullum (2002: 585), we adopt the premise that DCCs with *too* modifying a fear adjective entail the negative of the degree complement. This is suggested by their apparent uncancellability:

(4) The woman's grandchildren were in another room when the incident took place and were too afraid to intervene. (ZA, 2018, *News24*)

(4′) *The woman's grandchildren were in another room when the incident took place and were too afraid to intervene but intervened anyway.

This appears to be a consequence of the semantics of fear adjectives and the co-referentiality of the matrix and degree complement subjects, i.e. the uncancellability need not generalize to other categories of adjective.[3]

Within the category of degree complements, the present chapter employs a further terminological distinction based on the identity of the degree particle heading the DCC—degree complements dependent on *too* and those dependent on *so* are referred to as *too* complements and *so* complements, respectively. There is an important relationship between *too*-DCCs and *so*-DCCs—in accordance with the negative entailment of *too*-DCCs just mentioned, the two constructions form paraphrase pairs where an affirmative *too*-DCC corresponds to a *so*-DCC where the clause corresponding to the *too* complement is negative (Quirk et al. 1985: 1142). Also, in contradistinction with the *too*-DCC, the *so*-DCC requires its degree complement to be finite, with an overt subject that need not be co-referential with any matrix constituent. Thus, the *so*-DCC is more explicit than the *too*-DCC (Huddleston and Pullum 2002: 1211; Rohdenburg 2015: 103), requiring overt specification of semantic features that the *too*-DCC leaves implicit. In other words, semantic features that the *too*-DCC only implies—or leaves to be pragmatically inferred—are laid bare in a corresponding *so*-DCC. For example, (1a–b) and (2b) above can be paraphrased as follows:

(1a′)   Everyone quickly made it into classrooms, except for one little girl who was across the playground, so scared that she couldn't move.

(1b′)   Everyone's so scared they won't speak up about it so I do in a jokey passive aggressive way.

(2b′)   I'm so tired I can't proofread this wretched article.

As these examples illustrate, the *so* complement often features a modal auxiliary. There are at least two reasons. Firstly, though an unmodalized verb is not impossible, the semantic content of a simple *not*-negation is entailed by many negated modals. Using an appropriate negated modal can thus add semantic nuance at little to no cost in phonological complexity—*X couldn't Verb* entails *X didn't Verb* in the same number of syllables.

Secondly, fear typically affects the subject-referent's volition or ability to participate in the complement situation. Since volition and ability both fall into the domain of dynamic modality (Palmer 1990: 36; Huddleston and Pullum 2002: 178, 192), dynamic (uses of) modals such as *can* and *will* are a natural match for *so* complements involving adjectives of fear.

Another consequence of the lesser explicitness of infinitival clauses relative to finite ones is that the paraphrase relationship between an infinitival *too* complement and the corresponding finite *so* complement is one-to-many rather than one-to-one. A *too*-DCC involving a fear adjective communicates, imprecisely but economically, that the referent of the matrix subject has exceeded the upper limit of fear compatible with the actualization of the complement situation. This semantic content can often be expressed by several different modals—for example, the *so* complement in (1b′) could alternately have *daren't*. Note, also, that the choice is not restricted to central modals—*couldn't* and *won't* in (1a′–b) could be replaced by the lexical modal expressions *was unable to* in the former case and *refuse to* in the latter, or by any other pair of lexical modal expressions conveying the appropriate degree of inability and unwillingness, respectively. Conversely, a *so*-DCC whose matrix and degree complement subjects are co-referential can always be rendered by a *too*-DCC, often at a minor loss of semantic information:

(4) It is possible to think that gay people don't exist because gay people are so frightened that they daren't say they exist. (CA, 2017, *ETCanada.com*)

(4′) It is possible to think that gay people don't exist because gay people are too frightened to say they exist.

(5) "The family are decent people but they're so afraid that they're not assisting gardai over fear of reprisal attacks." (IE, 2017, *Herald.ie*)

(5′) "The family are decent people but they're too afraid to assist gardai over fear of reprisal attacks."

(6) "It's all to get people so scared that they don't question what's going on in their own country," he said. (CA, 2017, *Times Colonist*)

(6′) "It's all to get people too scared to question what's going on in their own country," he said.

In all three examples, the *too*-DCC paraphrase is a more compact rendition of the original—in (4′) the explicit modal *dare* is lost in the paraphrase, but *too frightened to* implies it sufficiently. In (5′), the *too*-DCC paraphrase does away with the aspectual distinction provided by the progressive construction, but it is hard to envision a context where this would impede communication. Example (6) features a simple negation without a modal, and there seems to be no information loss involved in the paraphrase.

Finally, regarding negation, it is useful to generalize the polarity relationship between a *too*-DCC and the corresponding *so*-DCC as a matter of reversal—the polarity of a *too* complement is the opposite of the corresponding *so* complement, and vice versa. This is illustrated by (7) below:

(7)   "The Government is in cohorts with the banks and have created an environment where people are too terrified not to pay their mortgage despite the fact they might be paying 90% of their wages over..." (IE, 2019, *Irish Mirror*)

(7′)   The Government is in cohorts with the banks and have created an environment where people are so terrified that they pay their mortgage despite the fact they might be paying 90% of their wages over.

Coupled with its paraphrase relationship with the *too*-DCC, the explicitness of the *so*-DCC makes it a useful diagnostic tool in the analysis of the former construction. This will become more evident in later sections.

3.2 DATA AND METHOD

All data examples in the present chapter come from the NOW Corpus. Maintained by Mark Davies at the Brigham Young University, it is a dynamically growing database of English-language news from 20 countries in which English is either the native language or an official language. This corpus was chosen both due to its great size (over 5 billion words) and because we had access to a static local copy, enabling complex regular-expression searches to be carried out in a large sub-section of the corpus. We first consulted www.thesaurus.com to identify synonyms of *afraid*, then ran preliminary searches of the web version of the NOW Corpus to identify the ones whose non-finite alternation was chiefly limited to the *to* infinitive and *of* -*ing*. Eight such adjectives were identified: *afraid, scared, terrified, fearful, frightened, petrified, apprehensive,* and *timorous*. To explore the co-occurrence patterns of fear complements and *too* complements, we then searched a part-of-speech tagged local copy of the NOW Corpus that covers the time period from January 2010 through October 2017 and comprises 5 billion words.

Table 3.1 Raw and normalized frequencies of non-finite fear complements of five fear adjectives within a superordinate *too*-DCC with an overt degree complement in a 5-billion-word section of the NOW Corpus

Adjective	To *infinitives*	Of -ing	Total	Per billion words
afraid	16	34	50	10
scared	17	22	39	8
terrified	1	3	4	1
fearful	0	3	3	1
frightened	4	8	12	2
Total	38	70	108	22

Our regular expression (see Appendix) targeted all sentences that might possibly feature a fear complement of one of the eight adjectives nested within a *too*-DCC in which the degree complement is present. As seen in Table 3.1, the nesting of non-finite fear complements within degree complements headed by *too* is rare. *Afraid*, which seems to be the most common fear adjective in terms of DCC-nested fear complements, contributed only one observation per 100 million words. *Petrified, apprehensive*, and *timorous* all failed to yield a single DCC-nested complement of their own.

The remainder of this chapter discusses issues related to the nesting, omission, co-occurrence, and identification of fear complements, *too* complements, and infinitival adjuncts in the context of *too*-DCCs involving one of the five adjectives of fear listed above.

3.3 Four Types of Non-finite Complement

A *too* complement may be omitted if its semantic content is sufficiently clear from the context (Quirk et al. 1985: 1140; Biber et al. 1999: 528; Huddleston and Pullum 2002: 585). This is seen in (8a–b):

(8a) Meanwhile, Ariana previously admitted she wants to go tour but is too scared. (NZ, 2018, *New Zealand Herald*)

(8b) "It keeps you from doing things you otherwise might have because you're too terrified." (US, 2014, *People Magazine*)

The omitted *too* complement in (8a) is interpretable as the ellipsis of the complement of *want* in the superordinate clause, i.e. *to go tour*. In (8b) the *too* complement is also omitted, but its content is semantically present in the prepositional object of the matrix clause, viz *doing things you otherwise might have*.

Crucially, fear adjectives are perfectly capable of taking non-finite complements of their own, i.e. fear complements. Fear complements can be infinitival or gerundial, as illustrated in (9a–b) below. Also note, as seen in (9c), that fear complements are every bit as optional as the *too* complement:

(9a)   "Children who already have asthma should not be frightened to exercise because of their condition." (IE, 2018, *The Irish Sun*)

(9b)   Fawzy said he's terrified of being kidnapped and spirited to Egypt. (CA, 2012, *National Post*)

(9c)   He said to me, 'How many times have you stopped over the edge of a diving board and not jumped because you're scared?' (AU, 2010, *Sydney Star Observer*)

In particular, it is possible for an infinitival fear complement to appear within a DCC in which the *too* complement is omitted, and it is possible for both complements to co-occur. It is likewise possible for a gerundial fear complement to occur within a DCC. These patterns are illustrated by (10a–c), respectively:

(10a)   The problem in countries such as the UK is that people shy away from an entrepreneurial mindset because they're too scared to fail. (GB, 2014, *Evening Standard*)

(10b)   ... in smaller German towns, gays and lesbians are practically invisible and are too afraid to be recognised by neighbours or bosses to join a gay parade ... (GB, 2014, *The Guardian*)

(10c)   Doctors say the true toll may be much higher because wounded minority Uzbeks are too afraid of being attacked again to go to hospitals. (CA, 2010, *Toronto Star*)

In (10a) the need, purpose, or desire, i.e. exercise of entrepreneurialism, is identified by the matrix clause, so an explicit degree complement is unnecessary. The source of fear, by contrast, is not similarly clear from the context, so it is made explicit by a fear complement. Though examples like (10a) have no degree complement, the sentence can be reformulated as a DCC headed by *too* or *so* by encoding the stated need, purpose, or desire into a degree complement and, in the latter case, reversing the polarity. This is seen in (10a′) and (10a″) below, in the latter of which *won't* could be replaced by another compatible modal or even by an unmodalized negated VP:

(10a′)   The problem in countries such as the UK is that people are too scared to fail to embrace an entrepreneurial mindset.

(10a″)   The problem in countries such as the UK is that people are so scared to fail that they won't embrace an entrepreneurial mindset.

In (10b–c), both complements are present. The degree complement in (10b) is *to join a gay parade*, and the source of fear presented as an impediment to it is the prospect of being seen by neighbors or bosses, expressed by the fear complement. In (10c), the fear complement is morphosyntactically distinct from the degree complement. Though the frequency distribution of infinitival and gerundial fear complements appears to be linked to certain syntactic and semantic features (Vosberg 2003: 308; Rudanko 2017: 20; Chapters 2 and 4 of this volume), the substitutability of infinitival fear complements by *of -ing* is another key feature distinguishing them from degree complements.

Finally, an infinitival adjunct of result or higher purpose may follow the degree complement in order to elaborate its meaning. This is seen in (11):

(11)   Manoj Narra is one of the students facing deportation and says he is struggling to sleep and too scared to go home to face his family. (NZ, 2016, *Radio New Zealand*)

The source of fear, left implicit by the omission of the fear complement, is here the expected reaction of the family upon hearing the bad news. *To go home* is a degree complement—it is the element that becomes finite and undergoes a reversal of polarity in a semantically equivalent *so*-DCC. These roles are well illustrated by a paraphrase that converts *to face his family* into a fear complement, as seen in (11′):

(11′)   Manoj Narra is one of the students facing deportation and says he is struggling to sleep and so scared to face his family that he dare not go home.

Though a paraphrase like (11′) enables the degree complement to be identified, it takes the liberty of reordering the two clauses. This does not help determine the syntactic status of *to face his family* in (11), where it follows the degree complement rather than preceding it. Syntactically more helpful than (11′), therefore, is a *so*-DCC paraphrase like (11″), which retains the original ordering:

(11″)   Manoj Narra is one of the students facing deportation and says he is struggling to sleep and so scared that he dare not go home to face his family.

Although *to face his family* in (11) alludes semantically to the source of fear, it is syntactically separated from the fear adjective by a higher-level constituent, so cannot be a (syntactic) fear complement. It is, moreover, not substitutable by *of -ing* like a fear complement would be. This is evidenced by the ungrammaticality of (11‴):

(11‴)   *Manoj Narra is one of the students facing deportation and says he is struggling to sleep and so scared that he dare not go home of facing his family.

To face his family is therefore neither a degree complement nor a fear complement. Since no other elements capable of taking a complement clause are present, it must be an adjunct.

To recapitulate, DCCs headed by *too* exhibit at least four distinct patterns of infinitival clauses: (a) degree complement only, (b) fear complement only, (c) fear complement + degree complement,[4] and (d) degree complement + adjunct, with each combination illustrated in turn by (12a–d) below:

(12a)   He was too petrified to run away as they smashed the windows of his tempo. (IN, 2017, *Scroll.in*)

(12b) White people can never really address the deep systemic, perva-
sive problem of racism, because they're too afraid to be called
racist. (CA, 2018, *Lynn Journal*)

(12c) U.S. politicians are too afraid to alienate their voting base
to reach any substantive agreement on anything. (US, 2011,
TIM)

(12d) "Some people are too scared to even open their doors to go
out and buy food," he added. (IN, 2017, *The New Indian
Express*)

The next section discusses the syntactic and semantic ambiguities that may
arise as a result of this multiplicity of possible combinations.

3.4 Syntactic and Semantic Indeterminacy

The optionality and mutual independence of fear complements and
degree complements has the consequence that when an adjective of fear
modified by *too* is followed by only one *to* infinitive, it can be difficult to
identify whether its licensor is *too* or the adjective. Example (13) is such
a case:

(13) Sees waved for his partner, Alvin Tables, Jr. to swim out to the
boat. The girl, too afraid to swim, refused to go to the boat.
(GB, 2017, *FL Keys News*)

There are two plausible syntactic analyses of such examples. On the
degree-complement reading, *to swim* is licensed by *too*. There is then no
overt fear complement, and a more explicit paraphrase using a *so*-DCC
produces (13'):

(13') The girl, so afraid that she dared not swim, refused to go to
the boat.

On the alternative reading, the *to* infinitive is a fear complement, i.e. a
complement of *afraid*. There is no degree complement, but the semantic
content normally encoded by a degree complement is given by *to go to the
boat* in the main predicate. Also note that this main predicate is (lexically)

negated, making its semantic properties an exact match for a prototypical *so* complement of a fear adjective. This analysis yields the *so*-DCC paraphrase seen in (13″):

(13″)   The girl was so afraid to swim that she refused to go to the boat.

Which analysis is correct? Arguably, (13) exemplifies what Leech and Coates call merger (1980: 81–82). Though two distinct and viable analyses exist, their semantic difference is rendered inconsequential by the context. Note that the *so*-DCC paraphrases corresponding to the two rival analyses, respectively, are virtually synonymous. Though we assume that the author intended one or the other syntactic construction, the interpretive choice between them is of little semantic consequence in this context.

Regarding the identification of the syntactic roles of two infinitival constituents in a *too*-DCC, examples (14–16) lie on a cline of ascending analytical difficulty:

(14)   Are Catholic organisations too afraid to be labelled lefties to really criticise the wealthy and powerful? (AU, 2017, *Eureka Street*)
(15)   One mother said she was now too frightened to walk down to the Co-op in the village to buy a loaf of bread. (GB, 2017, *The Guardian*)
(16)   The victims will be too scared to go to line up to point out these monsters. (JM, 2017, *Jamaica Observer*)

Example (14) resolves easily due to the time relations. The criticism of the wealthy and powerful is presented as a potential <u>cause</u> of derogatory labeling, not as its effect. It thus cannot be an adjunct of result or higher purpose, so must be a *too* complement, i.e. degree complement, and the infinitive prior to it must be a fear complement.

Example (15) is less straightforward, since the time relations fail to rule out an adjunctive reading of the second infinitival. However, close examination suggests *to buy a loaf of bread* must be an adjunct. If it were a degree complement, the entailment would be that she was unable to buy a loaf of bread. This would violate the first Gricean maxim of quantity

(Grice 1975: 45). If a fear of walking to the store prevented her from buying bread, surely it would prevent her from buying other groceries at the Co-op as well—indeed any groceries whatsoever. If this was the case, then why stop at only mentioning bread? Surely the fear would not be limited exclusively to bread-acquisition outings, without an analogous effect on similar excursions made for milk and vegetables? More plausibly, the main clause of the degree complement is *to walk down to the Co-op in the village*, with the second infinitive serving as an adjunct. Such an analysis makes the *bread* clause easier to account for—its role is not to convey a generalized inability to purchase bread, but to highlight the speaker's loss of freedom to undertake spontaneous store trips. Accordingly, a *so*-DCC paraphrase such as (15′) below, with the *bread* clause as a degree complement, is implausible. More plausible is an analysis where the *bread* clause is an adjunct, and categorical incapacitation from buying bread is not entailed. Paraphrased as a *so*-DCC, this analysis yields (15″):

(15′)   One mother said she was now so frightened to walk down to the Co-op in the village that she couldn't buy a loaf of bread.

(15″)   One mother said she was now so frightened that she couldn't walk down to the Co-op in the village to buy a loaf of bread.

All of this suggests that in (15), *to buy a loaf of bread* is dominated by *to walk down to the Co-op in the village* and is therefore its postmodifying adjunct, not a higher-level constituent such as a degree complement. The same conclusion can be reached by substituting *of -ing* for the hypothetical fear complement. Doing so results in (15‴) which, while grammatical, either carries the same implausible negative entailment as (15′) or is missing a degree complement:[5]

(15‴)   One mother said she was now too frightened of walking down to the Co-op in the village to buy a loaf of bread.

Example (16) is trickier still, because the close association between the meanings of the two infinitival clauses may add a layer of pragmatic ambiguity. As regards the structure of the sentence, there are two plausible analyses. If *to go to line up* is a fear complement, then *to point out these monsters* is in all likelihood a degree complement, yielding the categorical entailment that the victims cannot identify the perpetrators. If, on the other hand, *to go to line up* is not a fear complement, then it is the first part of a degree complement within which *to point out these monsters* is an adjunct. This makes the entailment less general, negating only the victims' ability to go to a line-up to identify the perpetrators. However, there seems to be a close association between going to a line-up and identifying a perpetrator. At the same time, this association does not amount to a truth condition—going to a line-up to point them out is not the only way to identify perpetrators. Therefore, from a strictly semantic perspective, the structure in which *to point out these monsters* is an adjunct does not amount to a categorical negation of the victims' ability to identify the perpetrators. It only entails that the victims cannot identify the perpetrators through a procedure that requires them to go to a line-up—"local" rather than "global" incapacitation. It fails to cover options such as photographic line-ups presented to the victims at their homes. However, if the association between going to a line-up and identifying a perpetrator is very strong, global incapacitation, e.g. 'they cannot point out the perpetrators' could be the intended implicature even though it falls short of a truth-conditional entailment. Paraphrase (16′) illustrates the reading on which *to point out these monsters* is an adjunct, entailing only local incapacitation, i.e. inability to point out a perpetrator in a line-up requiring travel. Global incapacitation may be overlaid as an implicature but is not entailed. By contrast, paraphrases (16″) and (16‴) both posit a fear complement, entailing global incapacitation, i.e. inability to identify the perpetrator at all.

(16′)   The victims will be so scared that they cannot point out these monsters in a line-up that requires them to leave their houses.

(16″)   The victims will be so scared to go to a line-up that they cannot point out these monsters (at all).

(16‴)   The victims will be too scared of going to a line-up to point out these monsters (at all).

If *to point out these monsters* is an adjunct, that adjunct is a modifier of the negative finite complement of the *so*-DCC paraphrase, qualifying the semantic scope of what is negated. If it is a degree complement, its negation in any semantically accurate paraphrase is full and unqualified. The two analyses are semantically distinct, so this is not a case of merger. Rather, the undecidability emanates from the close semantic association between conventional line-ups and identification of perpetrators, which increases the probability that the one construction may be used to implicate what the other entails.[6]

To sum up the section, three criteria seem advantageous in identifying the syntactic role of each *to* infinitive when two of them follow a fear adjective premodified by *too*. If the situation denoted by the first infinitive is conceptualized as subsequent to that of the second infinitive, the first infinitive is likely a fear complement. When these time relations do not hold, further insight is gained by applying *so*-DCC paraphrases and gerundial substitutions that preserve the order of the infinitival clauses. Then, a clause that turns finite with its polarity reversed in a meaning-preserving *so*-DCC is a degree complement. A clause that can be replaced by a prepositional gerund while preserving grammaticality and the original meaning is a fear complement. An infinitival clause whose semantic content, but not necessarily syntactic form, remains as a qualifying element in the resulting paraphrase is an adjunct.

3.5 Conclusion

This work arose when we were looking into non-finite complementation of fear adjectives. We noticed that sometimes the adjective was modified by *too*, yet the *to* infinitive that followed did not seem a degree complement but appeared to be licensed by the adjective itself. Our corpus searches confirmed that fear complements and *too* complements were able to co-occur, corroborating our suspicion that a *to* infinitive following a fear adjective preceded by *too* could not automatically be classified as a degree complement. Despite the attention paid to *to* infinitive complements in the literature, their nesting in DCC contexts appears to have been overlooked till now. The search also yielded instances of *too* complements followed by infinitival adjuncts, as well as combinations where the identification of the syntactic role of each infinitival clause proved an arduous enterprise. We have now proposed three criteria—one concerning the time relations and two employing syntactic transformations—to help ascertain the syntactic relationships. This has clarified our own thinking of the phenomena concerned, and we hope that they may also prove helpful to others.

Clear cases of a *too*-DCC with an embedded infinitival fear complement are rare. Though we have hopefully succeeded in basing the present analysis on corpus examples representative of Standard English, those examples were hard to come by. It is also worth noting that, as shown in Sect. 3.2, there seem to be more gerundial fear complements than infinitival ones in DCC contexts. Examples are seen below:

(17)   Mrs Thompson used to be too afraid of getting stuck in an aeroplane seat to go on foreign holidays. (GB, 2013, *Daily Mail*)

(18)   He is the equivalent of a washed-up high school quarterback who is too scared of deflating his own ego to leave his small town. (US, 2016, *MTV.com*)

A majority of gerundials is contrary to the expected behavior of fear complements. According to recent work, the overall tendency is for fear adjectives to prefer *to* infinitives (Rickman and Rudanko 2018: 15–74). If fear complements within *too*-DCCs deviate from the overall selection pattern of fear complements, it will not have escaped notice that, unless omitted, the *too* complement is invariably introduced by *to*. Is this evidence of an *anticipatory horror aequi*? If it is, does the magnitude of the effect vary as a function of the distance between the fear complement and the *too* complement? Or is the operative factor simply the presence or absence of a superordinate DCC, reflecting Wasowian production planning considerations (Wasow 2002), where speakers' awareness of the imminent *to* infinitive—a consequence of the fixed structure of the DCC—steers usage away from infinitival fear complements within this construction, no matter the distance between the complements? These seem potentially rewarding questions for future research.

Finally, it may be worthwhile to study the role of prosody in influencing syntactic interpretations. According to Huddleston and Pullum, a strong prosodic boundary can overrule the default syntactic interpretation in favor of a marked structure when assigning the scope of negation (2002: 183). In addition, there exists some experimental evidence that prosody can influence hearers' interpretations where more than one syntactic reading is available (Price et al. 1991). With ambiguous examples such as (13) and (16) above, it seems plausible that even a fairly modest prosodic reinforcement may sway the hearer's syntactic interpretation in favor of one of the available analyses. Confirming or disconfirming such effects suggests itself as another worthy object of future study.

NOTES

1. Henceforth referred to as the NOW Corpus.
2. The derivational properties of degree complements have been discussed in Baltin (2006) and are not the focus of the present chapter.
3. Quirk et al. (1985: 1142) equate *it was too pleasant a day to go to school* with the cancellable 'it was such a pleasant day that I didn't want to go to school'. For their part, Huddleston and Pullum (2002: 1262–1263) translate the negation conveyed by *it is too late for you to go out now* using the modals 'can't or shouldn't', the latter of which they elsewhere characterize as cancellable (2002: 186). By contrast, *she was too tired to continue* features co-referential matrix and complement subjects, and Huddleston and Pullum analyze it as negatively entailing (2002: 585).
4. An infinitival adjunct could technically be added as a third *to* infinitive, though we would expect *horror aequi* to militate heavily against such constructions (Rohdenburg 2003: 236). Accordingly, we have found no corpus examples.
5. Our search of the 5-billion-word NOW Corpus did not return a single unambiguous instance where the degree complement was omitted and the adjective had a non-finite complement with a postmodifying infinitival adjunct. That combination, while conceivable, seems to be even rarer than the nesting of fear complements within DCCs.
6. There is also a third possible reading of (16), namely one in which *line up* is a VP rather than an anarthrous NP. However, we consider a VP analysis of *line up* to be implausible since it suggests that it is the victims who line up, rather than the suspects.

APPENDIX

(?i) too_\S+ (afraid|scared|terrified|apprehensive|fearful|petrified|
timorous|frightened)_\S+ ([^_]+_[rx]\S+)?(?#optional adverb)(to_\S+
([^_]+_[rx]\S+){0,2}(?#up to 2 optional adverbs)[^_]+_\S*v\S+|of_\S+
([^_]+_[rx]\S+){0,2}(?#up to 2 optional adverbs)\w+ing_\S+)
((?![.?!]_\S+ |[^_]+_(cs\S*|ccb\S*))(?# no intervening terminal
punctuation or subordinators allowed)\S+)*?to_\S+ ([^_]+_r\S+
){0,3}[^_]+_\S*v\S+\s

REFERENCES

Baltin, Mark. 2006. Extraposition. In *The Blackwell Companion to Syntax*, vol. II, ed. Martin Everaert and Henk van Riemsdijk, 237–271. Malden, MA: Blackwell.

Biber, Douglas, Stig Johansson, Geoffrey Leech, Susan Conrad, and Edward Finegan. 1999. *Longman Grammar of Spoken and Written English*. London: Longman.

Goldberg, Adele E. 1995. *Constructions: A Construction Grammar Approach to Argument Structure*. Chicago: University of Chicago Press.

Huddleston, Rodney D., and Geoffrey K. Pullum. 2002. *The Cambridge Grammar of the English Language*. 3rd printing 2010. Cambridge: Cambridge University Press.

Grice, Herbert P. 1975. Logic and Conversation. In *Speech Acts*, ed. Peter Cole and Jerry L. Morgan, 41–58. New York, NY: Academic Press.

Leech, Geoffrey, and Jennifer Coates. 1980 Semantic Indeterminacy and the Modals. In *Studies in English linguistics: For Randolph Quirk*, ed. Sidney Greenbaum, Geoffrey Leech, and Jan Svartvik, 79–90. London: Longman.

Palmer, Frank R. 1990. *Modality and the English Modals*, 2nd ed. London: Longman.

Price, P., M. Ostendorf, S. Shattuck-Hufnagel, and C. Fong. 1991. The Use of Prosody in Syntactic Disambiguation. *Journal of the Acoustical Society of America* 90: 2956–2970.

Quirk, Randolph, Sidney Greenbaum, Geoffrey Leech, and Jan Svartvik. 1985. *A Comprehensive Grammar of the English Language*. London: Longman.

Rickman, Paul, and Juhani Rudanko. 2018. *Corpus-Based Studies on Non-finite Complements in Recent English*. Houndmills, Basingstoke: Palgrave Macmillan.

Rohdenburg, Günter 2003. Cognitive Complexity and *Horror Aequi* as Factors Determining the Use of Interrogative Clause Linkers in English. In *Determinants of Grammatical Variation in English*, ed. Günter Rohdenburg and Britta Mondorf, 205–249. Berlin and New York: Mouton de Gruyter.

Rohdenburg, Günter. 2015. The Embedded Negation Constraint and the Choice Between More or Less Explicit Clausal Structures in English. In *Perspectives on Complementation: Structure, Variation and Boundaries*, ed. Höglund et al., 101–127. Houndmills, Basingstoke, Hampshire and New York, NY: Palgrave Macmillan.

Rudanko, Juhani. 2017. *Infinitives and Gerunds in Recent English: Studies on Non-finite Complements with Data from Large Corpora*. New York: Palgrave Macmillan.

Wasow, Thomas. 2002. *Postverbal Behavior*. Stanford, CA: CSLI Publications.

Vosberg, Uwe. 2003. The Role of Extractions and *Horror Aequi* in the Evolution of -*ing* Complements in Modern English. In *Determinants of Grammatical*

Variation in English, ed. Günter Rohdenburg and Britta Mondorf, 305–327. Berlin and New York: Mouton de Gruyter.

SOFTWARE USED

Anthony, Lawrence. 2018. AntConc (Version 3.5.7) [Computer Software]. Tokyo, Japan: Waseda University. Available from http://www.laurenceanth ony.net/software.

CORPORA CONSULTED

Davies, Mark. 2013 *Corpus of News on the Web (NOW): 3+ Billion Words from 20 Countries, Updated Every Day.* Available online at https://corpus.byu.edu/ now/.

CHAPTER 4

Semantics, Syntax, and *Horror Aequi* as Predictors of Non-finite Alternation: A Multivariate Analysis of Clausal Complements of *Afraid* in the NOW Corpus

Abstract Semantic, syntactic, and *horror aequi* factors are underresearched in the context of non-finite alternation. In particular, they have not been adequately investigated with modern multivariate methodology. The present study addresses this gap through multivariate analysis of non-finite alternation in complement clauses of *afraid* featuring covert subjects. The authors use purposeful sampling to construct a dataset enabling statistical inference on nine explanatory factors that have been proposed in previous literature. The factors fall into the three superordinate categories representing syntactic complexity factors, semantic factors, and *horror aequi* factors, respectively. Results of mixed-effects logistic regression suggest that Rohdenburg's Complexity Principle is valid except where it clashes with other predictive principles. Within the semantic domain, the Choice Principle and negation of the adjectival head are both found to exert a dramatic influence on variant choice, but the two effects do not seem to be additive. Retrospective *horror aequi* effects appear to vary significantly in magnitude as a function of the phonological complexity of the repeated construction. No evidence is found for anticipatory *horror aequi*.

Keywords Syntax · Non-finite complementation · Variation · Multivariate analysis

© The Author(s) 2021   69
J. Ruohonen and J. Rudanko, *Infinitival vs Gerundial Complementation with Afraid, Accustomed, and Prone,*
https://doi.org/10.1007/978-3-030-56758-3_4

4.1   INTRODUCTION

Consider the sentences in (1a–b), from the *Corpus of News on the Web* (Davies 2013):

(1a)   She was afraid to open her mailbox. (CA, 2014, *National Post*)
(1b)   I was afraid of staying here ... (NZ, 2017, *New Zealand Herald*)

In sentence (1a) the adjective *afraid* selects a *to* infinitive complement, and in (1b) the same adjective selects what has been called an *of -ing* complement, with the *-ing* constituent being a gerund. It is assumed here that both types of complement are sentential, and that the lower sentences have their own understood or implicit subjects. These assumptions are not shared by all linguists, but they make it possible to represent the argument structures of the lower verbs in a straightforward fashion. Further, the presence of understood subjects in nonfinite complements can be motivated on the basis of binding theory. As noted in Chapter 1, in a sentence of the type of *Perjuring himself would not worry John*, an understood lower subject is needed to bind the reflexive *himself*. Since understood subjects are therefore independently needed in the grammar of English, it is likewise reasonable to make provision for them in (1a–b).

Regarding the nature of the sentential complements in (1a–b) it is also clear that because the higher subjects in both (1a) and (1b) receive their theta roles from the adjective *afraid*, both types of sentences involve control (rather than NP Movement). The understood subjects can then be represented with the symbol PRO, in line with current work. The sentences (1a–b) are similar in that both involve subject control, with the higher subject determining the reference of PRO in each case. The constituent structures of (1a–b) may then be bracketed in their relevant aspects as in (1a′) and (1b′), with structure (1b′) also incorporating the convenient assumption that a gerund forms part of a nominal clause, that is, a sentence dominated by a NP.

(1a′)   [[She]$_{NP}$ was   afraid   [[PRO]$_{NP}$ [to]$_{Aux}$ [open   her   mailbox]$_{VP}$]$_{S2}$]$_{S1}$
(1b′)   [[I]$_{NP}$ was   afraid   [[of]$_{Prep}$ [[[PRO]$_{NP}$ [staying   here]$_{VP}$]$_{S2}$]$_{NP}$]$_{PP}$]$_{S1}$

The variation found in infinitival and gerundial complements of the adjective *afraid* and of other, generally less frequent adjectives expressing fear is well known in the literature, and it illustrates a case in point where one and the same head selects two different types of sentential complements that are close to each other in meaning, but still very different in their syntax. Such variation, occurring in the core grammar of English, is then especially intriguing because of Bolinger's Principle, which states that a "difference in syntactic form always spells a difference in meaning" (Bolinger 1968: 127).

In recent work new principles have been proposed to account for the variation between non-finite complements of the type found in (1a–b). Such principles have often emerged from the study of large corpora and a large database is also used here. In general terms it is the purpose of this chapter to consider such principles and to compare their impact using methods of advanced statistics. The comparison concerns current English, and it is undertaken on the basis of the *Corpus of News on the Web*, henceforth referred to as the NOW Corpus. In the remainder of this Introduction the principles in question are introduced and they are then examined and compared in Sect. 4.4.

One set of potentially explanatory factors falls under the umbrella of the Complexity Principle. First proposed by Rohdenburg, it is often stated as follows:

> In the case of more or less explicit grammatical options the more explicit one(s) will tend to be favoured in cognitively more complex environments. (Rohdenburg 1996: 151)

The same author (2015: 103) equates the notion of explicitness with that of clausal finiteness as defined by Givón, i.e. the degree to which the clause shares its characteristics with a prototypical transitive main clause (1990: 853).

Of the many complexity factors that have been proposed in previous literature (see Rohdenburg 2015: 122n for a list), one that serves as a major focus in the present study is the Extraction Principle. This principle emerged in the late 1990s and early 2000s, with pioneering work done by Günter Rohdenburg and Uwe Vosberg. The latter provided the following definition:

In the case of infinitival or gerundial complement options, the infinitive will tend to be favored in environments where a complement of the subordinate clause is extracted (by topicalization, relativization, comparativization, or interrogation etc.) from its original position and crosses clause boundaries. (Vosberg 2003a: 308; see also Vosberg 2003b)

In other work, including Vosberg (2006) and Rudanko (2006), the Extraction Principle has been extended to also include the extraction of adjuncts in cases where the extraction site can be determined (cp. Vosberg 2006: 69–70). Examples (2a–b) below feature the extraction of a complement and of an adjunct, respectively:

(2a)   Daniel Igali, man of hope and Canadian Olympic champion, has made a promise he is not afraid to share. (CA, 2013, *Globe and Mail*)

(2b)   Andrew Mason – with his mild brown hair, scruffy beard, and faded polo shirts – is funny in a way that most CEOs would be afraid to be. (US, 2011, *Gawker*)

The gap in (2a) corresponds to a direct object, whilst the one in (2b) corresponds to a manner adjunct.

Additional complexity factors taken into consideration in the present chapter include structural discontinuities, complement negation, and passivization. They are illustrated in (3a–d):

(3a)   They were afraid even to speak to one another. (GB, 2017, *Daily Mail*)

(3b)   Conversations were hushed, as though people were afraid of further angering the heavens. (PH, 2013, *Inquirer.net*)

(3c)   And if you are afraid of not having enough space for your bag, you know, that's not a good start for a nice flight experience. (AU, 2016, *The Australian Financial Review*)

(3d)   Nina Dobrev is not afraid to be photographed without makeup (SG, 2017, www.MICEtimes.asia)

Both (3a) and (3b) feature the insertion of an adverb before the head of the lower VP. However, (3a) exhibits what we will call pre-complement insertion, with the adverb preceding the complement clause, while (3b) exemplifies complement-internal insertion of an adverb. We are aware of no previous quantitative study distinguishing between the two types in the context of non-finite alternation. Therefore, both detection and comparison of the potential effects of these two types of insertion are among the objectives of this chapter.

Thirdly, (3c) features a non-finite complement that is negated. Previous research by Rohdenburg (2015) and Hagemeier (2006) has established univariate correlations between embedded negation and preference for more explicit syntactic variants, but their studies involve finite complements. The present work investigates whether their findings generalize to the alternation between the *to* infinitive and *of -ing*.

Fourth, the passive voice has been determined by psycholinguistic experiments to be cognitively more demanding than the active (Givón 1990: 958). Thus, the Complexity Principle predicts that passive subordinate clauses will favor the infinitival option, as seen in (3d).

A second set of syntactic factors potentially linked to variant selection falls under the umbrella of *horror aequi*. As another principle attributable to Rohdenburg, *horror aequi* has been defined as follows:

The *horror aequi* principle involves the widespread (and presumably universal) tendency to avoid the use of formally (near-) identical and (near-) adjacent (non-coordinate) grammatical elements or structures. (Rohdenburg 2003: 236)

Examples consonant with *horror aequi* are seen in (4a–b), where the precedence of *afraid* by a *to* infinitive and *of -ing*, respectively, presumably contributes to the selection of the opposite variant in the lower clause:

(4a) "We don't have to be afraid of doing everything we can to protect football." (GB, 2016, *WalesOnline*)

(4b) The idea of being afraid to move in implies men don't want to get married and that's wholeheartedly incorrect. (US, 2010, *CNN*)

Conceivably attributable to *horror aequi* is also the potential tendency of adjectives, unaddressed in previous literature, to favor the gerundial complement when nested within a superordinate degree-complement construction (DCC) headed by *too*. The *too*-DCC[1] licenses a *to* infinitive of its own, i.e. a degree complement, which is obligatory unless clear from the context (Quirk et al. 1985: 1140). Speakers' anticipation of the degree complement might contribute to a dispreference for infinitival complementation of *afraid* when the adjective is embedded within a *too*-DCC. This is seen in (5a), whereas (5b) contradicts the hypothesis:

(5a)   Without union protection, most farm workers are too afraid of being fired to raise concerns about dangerous conditions, the UFCW says. (CA, 2010, *Tillsonburg News*)

(5b)   Are Catholic organisations too afraid to be labelled lefties to really criticise the wealthy and powerful? (AU, 2017, *Eureka Street*)

Note that the definition of *horror aequi* does not specify whether the proposed dispreference for (near-)identical structures is expected to operate in a strictly retrospective fashion, or also in anticipation of obligatory elements. The latter tendency, if confirmed, would be highly reminiscent of the kind of production planning considerations discussed at length by Wasow (2002), and clausal complementation of *afraid* within superordinate *too*-DCCs is an informative grammatical context in this regard. A confirmatory analysis of such effects is another important objective of the present chapter.

Turning to semantics, the present chapter investigates two semantic factors that have previously been invoked to explain non-finite alternation. The first of them is the Choice Principle. It has been formulated as follows:

In the case of infinitival and gerundial complement options at a time of considerable variation between the two patterns, the infinitive tends to be associated with [+Choice] contexts and the gerund with [−Choice] contexts. (Rudanko 2017: 20)

A [+Choice] context is then defined as a context where the lower subject, represented by PRO in (1a′–b′), is an Agent. This is the case in (1a) and in (1b). In sentences where the lower subject is not an Agent, the context is [−Choice], and it is also easy enough to find sentences of that type, as in (6a–b).

(6a)   They enjoy a certain status and are afraid to lose it, and there-
       fore prey on people of a lower status to elevate their own. (CA,
       2013, *Macleans.ca*)

(6b)   I read on Friday about an old lady who has kept herself in
       her own home because she is afraid of falling down when she
       ventures outside. (SG, 2017, *The Middle Ground*)

The examples in (1a) and (6b) are in line with the Choice Principle, but
those in (1b) and (6a) run counter to it, and it is one of the objectives
of this chapter to investigate whether it is a worthwhile generalization
bearing on complement choice.

It is particularly worthy of note that the principal function of the
passive is to demote the agent to an optional argument, usually with
the consequence that the patient is promoted to subject status (Shibatani
1985). This has the consequence that passive subjects are overwhelmingly non-agents, i.e. the context is usually [−Choice]. Accordingly, the
majority of passive lower clauses represent a context in which the Choice
Principle and the Complexity Principle yield opposite predictions. The
present chapter seeks to inquire into the repercussions of this conflict.

The second semantic factor investigated in the present work is negation of the higher i.e. matrix clause. This includes both *not*-negation and
no-negation (Tottie 1991), including raised (Horn 1978), but excludes
external negation (Huddleston and Pullum 2002: 182–183). These
features are illustrated in (7a–d):

(7a)   He told CPJ, "I am not afraid to stand up for the truth. I will
       not be silenced." (ZA, 2017, *News24*)

(7b)   The provision requires that we be protected so that we are never
       afraid to speak and question power. (ZA, 2016, *R News*)

(7c)   "We don't want anyone to be afraid to be in our city, or
       walk the streets or go to school," Mr Pappas said. (NZ, 2017,
       Stuff.co.nz)

(7d)   We can't be afraid to fight for our dignity. (IN, 2016, *Zee News*)

Not-negation occurs when the higher clause is directly modified by *not*,
as in (7a), or by its contracted variant *n't*. *No*-negation encompasses all
cases in which the higher clause is negated by a determiner or adverb
incorporating the fused *n*-element, as in (7b). Negative raising occurs

when the negation, whether by *not* or *n-*, belongs syntactically to the superordinate clause but applies semantically to the subordinate clause. This is seen in (7c), which is semantically equivalent to 'we want no one to be afraid to...' Sentence (7d) exemplifies external negation. Its scope is limited, both syntactically and semantically, to the superordinate clause. Thus *afraid* remains unnegated, and the sentence is not paraphrasable by 'we can/may be unafraid to ...', translating instead to something like 'it is unacceptable for us to be afraid to...' Chapter 2 detected a statistically significant effect for *not*-negation of the higher clause, conditional on [±Choice], in British but not in Canadian English, where the pattern was similar but the data too sparse to support reliable statistical inference. The present investigation should be able to provide a valuable corroboration or disconfirmation of those preliminary findings.

Lastly, though the present chapter is not primarily focused on intervarietal variation, the inner-circle vs outer/expanding-circle distinction will be taken into consideration as a potential confounding variable in the statistical analysis (Maier et al. 2012).

4.2   Corpus

The NOW Corpus is a dynamic collection of English-language online news maintained by Mark Davies at Brigham Young University. The corpus is updated daily by an automated script that retrieves URLs of English-language news articles from Google News and batch-downloads the text from the associated webpages. Then, after an automated boilerplate screening followed by part-of-speech tagging and lemmatization carried out by the CLAWS4 tagger (Garside and Smith 1997), the texts are incorporated into the existing NOW database. The stored articles are classified by date and country, with the latter information provided by Google's geolocation algorithm. A total of 20 English-speaking countries are represented. In order to facilitate the kind of particularized corpus searches necessary, the present authors used a static local copy of the NOW Corpus. Its format is seen in (1a″), where tokenization and part-of-speech annotation have been left intact:

(1a″)   She_pphs1   was_vbdz   afraid_jj   to_to   open_vvi   her_appge
mailbox_nn1 ._.

The corpus covers a time period from January 2010 through October 2017, comprising a total of 5 billion words.[2]

4.3 DATA COLLECTION

The dataset was collected following a procedure that has been characterized as purposive (Lavrakas 2008: 645), selective (Tagliamonte 2012: 201), or purposeful sampling. As described by Duan et al. (2015), "the essence of purposeful sampling is to select information-rich cases for the most effective use of limited resources." In the case at hand, purposeful sampling was instrumental in enabling the authors to amass sufficient data on rare grammatical features to make statistical inference about their effects possible without having to manually code tens of thousands of observations.

To form the statistical backbone of the dataset, the authors first collected a randomized sample of 250 *to*-infinitival and 250 *of* -*ing* tokens. This was felt necessary to ensure that both "outcomes" were sufficiently represented to avoid data bottlenecks (Hosmer et al. 2013: 407), since previous research suggested that the infinitival variant was three or four times as frequent as *of* -*ing* (see Chapter 2). Both random samples contained a very small number of irrelevant tokens, examples of which are seen in (8a–c):

(8a) "The ANCYL of today, I'm afraid to say, is a disgrace and an embarrassment," said Loubser. (ZA, 2012, *Moneyweb.co.za*)

(8b) Because no one who's putting in 50 or 60+ hours because they're afraid not to is going to stick out their neck... (US, 2012, *ITworld.com*)

(8c) When it comes to who is afraid of overbearing government, Republicans tend to take the biggest slice of the cake at 81 percent. (US, 2017, *TheBlaze.com*)

In (8a), the sequence *I'm afraid to say* is a parenthetical, lexicalized unit (Biber et al. 1999: 58–59), roughly equivalent in meaning to a stance adverb such as *regrettably*. Such multi-word lexical units show limited possibilities of substitution (ibid), so the *to* infinitive in (8a) is hardly replaceable by *of* -*ing*. Therefore, tokens like (8a) fall outside the "variable context" and must be excluded (Tagliamonte 2006: 86–88).

Similarly, the omission of the verbal head seen in (8b) is impossible with the gerundial variant, so the context is not variable. Lastly, in (8c), *overbearing government* is a NP, thus not constituting a clausal complement at all.

This sampling stage yielded 246 legitimate infinitival tokens and 244 gerundial ones, respectively. Next, regular expressions combining lexical and part-of-speech information with negative lookarounds and optional capture groups (Friedl 2002) were crafted to target rare grammatical features of interest. Due to their low overall frequency and privileged status in this study, extractions and tokens within *too*-DCCs were sampled to the full extent that the regular expressions targeting them yielded relevant results. To give one example, the regular expression used to capture non-finite complements of *afraid* nested within *too*-DCCs is seen in (9) below. Gray sequences of the form (?#text) are explanatory code comments.

(9)   (?i)\stoo_\S+\safraid_\S+\s([^_]+_[rx]\S+\s)?(?# optional adverb

or negator)((?# begin search for non-finite complement)

to_\S+\s([^_]+_[rx]\S+\s){0,2}(?# up to 2 optional adverbs

or negators)[^_]+_\S*v\S+|of_\S+\s([^_]+_[rx]\S+\s){0,2}

(?# up to 2 optional adverbs or negators)\w+ing_\S+

(?# end search for non-finite complement))\s((?![.?!]_\S+

\s|[^_]+_(cs\S*|ccb\S*))\S+\s)*?(?# any number of intervening

tokens as long as they are not terminal punctuation or

subordinators)to_\S+\s([^_]+_r\S+\s){0,3}[^_]+_\S*v\S+

(?# to VERB, with up to 3 optional adverbs inbetween)\s

This search had low precision due its open-ended nature. It returned a total of 246 hits, only 52 of which were relevant ones of the type exemplified by (5a–b). However, such exhaustive and open-ended searches are necessary with features as rare as this one unless the corpus is much larger still. Extractions, by contrast, are sufficiently frequent that they could be targeted with a query sacrificing recall for precision, yielding 170 relevant hits out of a total 327. In addition, 11 extraction cases were returned fortuitously by other searches. The additional grammatical features of interest, such as complement negation and insertions, were much easier to target with high precision due to their straightforward syntactic characteristics. 50 random observations of each additional syntactic feature were sampled in addition to all fortuitous tokens found in the batches not specifically targeting that feature. On a small number of occasions, a targeted search yielded a token that had already been fortuitously retrieved by a more general search, or vice versa. Such duplicates were removed manually. All regular expressions used in retrieving the data are provided in the Appendix. All randomization was done using R's (v3.4.4) random sampling and random number generation functions.

One downside of purposeful sampling is that, since rare features or rare outcomes (in our case, both) are oversampled, the resulting dataset is no longer representative of the population distribution of the two outcomes. However, since the two initial samples of 250 observations of each outcome were both randomly drawn from a pool of observations resulting from an exhaustive search of the corpus for that outcome, comparing the grand totals of the two exhaustive pools is likely a reasonable approximation of the true population proportion of the variants—at least in the News register. The grand total yielded by the exhaustive search for *to* infinitives in the corpus is 39,225 (77%), while that for *of* -*ing* is 11,622 (23%). This overall distribution is congruous with previous research on the phenomenon.

The separate sampling stages yielded a grand total of 1,392 tokens, of which 990 were found relevant and included in the ensuing statistical analysis.

4.4   ANALYSIS OF THE DATA

4.4.1   Model Selection

We used the *lme4* package in R to fit mixed-effects logistic regression models to the data.[3] An overview of this statistical method is given in Sect. 2.3.3. The initial model was as follows:

Fixed Effects:

1. Extraction (trichotomous: None, Complement, Adjunct)
2. Insertion (trichotomous: None, Pre-Complement, Within-Complement)
3. Complement negation (dichotomous)
4. Voice (dichotomous)
5. Immediate prior *aequi* (trichotomous: None, To Verb, Of Verbing)
6. Superordinate *too*-DCC (dichotomous)
7. [±Choice] (dichotomous)

8. Matrix negation (trichotomous: Unnegated, *Not*-Negation, *No*-Negation)
9. Non-inner-circle variety (dichotomous)

Random Effects:

1. Subordinate verb (nominal-scale with 245 categories)

From this starting point, we followed the stepwise model-selection procedure detailed in Sect. 2.3.4 above. We began with a routine check for multicollinearity (Fox and Weisberg 2019: 429–434), which detected no problems. As in Chapter 2, Voice turned out to be inconsequential in variant selection, with a coefficient of −0.1 and a *p*-value of 0.81.[4] Dropping this variable caused no confounding, so we removed it from the model. The same held true for the nesting of the complement within a DCC. The coefficient for DCC-nested complements was 0.34 i.e. favorable to the infinitive, which was contrary to our expectations. It was, moreover, statistically non-significant with a *p*-value of .37. We thus dropped this variable from the model after confirming that no confounding resulted.

Complement negation had a statistically non-significant odds ratio of $e^{0.55} = 1.74$ favoring *of* *-ing*[5] ($p = 0.19$). This is contrary to the Complexity Principle. Section 4.4.2 below proposes an explanation, but for model-selection purposes the variable was superfluous, so we removed it after checking for confounding.

The next candidate for removal was *to*-infinitival *aequi*, whose gerund-favoring coefficient of 0.31 fell far short of statistical significance ($p = .33$). We removed its indicator variable from the model after checking for confounding.

The effect of adjunct extractions could not be reliably estimated. The extractions in the dataset involved almost exclusively complements, with only seven data points featuring an adjunct gap. At 1.73, the coefficient reported for adjunct extraction was similar to the value of 1.48 reported for complement extraction, and merging the two categories had no perceptible impact on the model fit or on the remaining coefficients. We therefore proceeded with a simplified model where no difference was made between subtypes of extraction.

Not-negation and *no*-negation of the higher clause had similar coefficients at 0.74 and 1, respectively. Both were statistically significant, with $G = 14.78$ and $p < .001$ for *not*-negation and $G = 6.54$ and $p = .01$ for *no*-negation, respectively. This is a good pedagogical example of the difference between effect size and statistical significance, with the smaller coefficient more statistically significant than the larger one because its estimate is backed by more data (there were 290 *not*-negations but only 48 *no*-negations). In the interest of model parsimony and simplicity of interpretation (Bilder and Loughin 2015: 266), the two negation types were subsumed under a single superordinate category of matrix negation. The model fit was essentially unaffected by this merger ($G = .3$; $p = .54$).

At this point, all remaining predictors seemed to be both practically and statistically significant, so we proceeded to check for interactions between them. A very similar interaction was found between Choice and matrix negation as in the British data in Sect. 2.4.3, and it was very statistically significant ($p < .001$). We thus kept this interaction term in the model.

4.4.2   Model Interpretation

As intended, combining a larger dataset with the purposeful sampling approach allowed statistically significant results to be obtained on many more variables than in Chapter 2. The final model has a classification accuracy of .8 and a concordance index of .872. The model stands as follows:

Fixed Effects:

Complexity factors

1. Extraction (dichotomous)
2. Insertion(s) (trichotomous: None, Pre-Complement, or Within-Complement)

Aequi factors

3. Immediate precedence of *afraid* by *of -ing* (dichotomous)

Semantic factors

4. [±Choice] (dichotomous)
5. Matrix negation (dichotomous)

Interactions

6. Choice × Matrix negation (dichotomous)

Control variables

7. Non-inner-circle variety (dichotomous)

Random Effects:

8. Subordinate verb (nominal-scale with 245 categories)

Figure 4.1 presents the point estimates of all variables included in the final model, along with their 95% profile-likelihood confidence intervals (Hosmer et al. 2013: 15–20). The point estimates represent estimated effect sizes and are therefore the parameters of primary interest.

Confidence intervals are analogous to p-values, quantifying the degree of uncertainty around the point estimate. They are inversely proportional to effective sample size. A confidence interval extending to both sides of zero implies a p-value above .05, i.e. lack of statistical significance, in a likelihood ratio test. The Intercept has been omitted from the figure because the purposeful oversampling of *of -ing* complements and other marked features renders the in-sample baseline log odds of infinitival complementation strictly unrepresentative of the true population parameter and hence irrelevant from a linguistic perspective.

4.4.2.1   Complexity Effects
Extraction emerges as the strong infinitive-favoring factor that earlier work has predicted it to be. Complement clauses with a gap left by an extracted complement or adjunct are estimated to have $e^{1.49} = 4.45$ times the odds of selecting a *to* infinitive compared to analogous non-extraction contexts ($p < .001$). Likewise, the presence of a pre-complement insertion

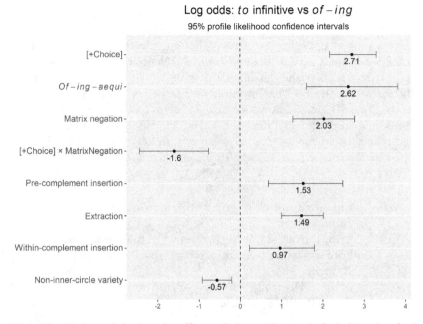

Fig. 4.1 Estimated logit-scale effects of the predictors included in the final model

exhibits a strong effect, with an estimated odds ratio of $e^{1.53} = 4.64$ ($p < .001$) compared to analogous tokens without such insertions. The effect of complement-internal insertions is less dramatic but still considerable at $e^{0.97} = 2.64$ ($p = .01$). This matches our expectations. We suspected that the effects of the two types of insertion might differ in magnitude because the first type occurs before the complement, thus potentially delaying the choice, while the latter occurs when the complement variant has already been selected, increasing syntactic complexity but not delaying the choice per se. However, both effects are broadly in line with the Complexity Principle, and the p-value for the trichotomous variable contrasting each type of insertion with contexts without any intervening material is below the .001 mark ($G = 18.58$; $df = 2$). It is also worth mentioning that the confidence intervals overlap, so conflating the insertion categories would be wholly justifiable from a strictly statistical perspective. However, we find the observed effect differential interesting, so we leave the categories separate.

The non-significant result on complement negation requires comment. This feature may constitute a syntactic context where two complexity factors pull in opposite directions. Though it may be true that the *to* infinitive is more explicit than *of -ing* (Rohdenburg 2016: 471–472), it is equally true that the default position of *not* in negated *to* infinitives increases complexity. The insertion of *not* before *to* increases the size of its Phrasal Combination Domain (Hawkins 2004), thereby violating the Minimize Domains Principle (ibid). More specifically, a sequence of the type *X is afraid not* ... has a larger number of syntactic paths still open than the type *X is afraid of* ... As an illustration, consider examples (10a–c):

(10a)   I'm afraid not all of you will be moving up. (GB, 2016, *The Sunday Times*)

(10b)   "No, I'm afraid not Michael, you're attempting to look at this from a perspective, not a prospective." (CA, 2014, *Good E-Reader*)

(10c)   Is the NHS really the "envy of the world"? No, I am afraid not. (GB, 2018, *Yorkshire Post*)

Instead of taking a non-finite complement, *afraid not* may alternatively be followed by a finite complement, as in (10a), a vocative NP, as in (10b), or it may even constitute the (elliptical) end of the sentence, as in (10c). The same multitude of options is not available after *afraid of*, which must be followed by a nominal or gerundial complement.[6] This domain minimization advantage in the specific context of negated complements may be a factor offsetting the general preference for *to* infinitives in complex syntactic environments.

4.4.2.2   Aequi *Effects*

The model shows a striking *horror aequi* effect against two *of -ing* constructions separated only by *afraid*. With other things equal, the odds of *afraid* selecting a *to* infinitive when immediately preceded by *of - ing* are estimated at $e^{2.62} = 13.7$ times the odds when not immediately preceded by *of -ing* ($p < .001$). No analogous effect is apparent with preceding *to* infinitives, whose statistically non-significant odds ratio is estimated at only $e^{0.32} = 1.37$. To be sure, the *of -ing* construction is phonologically heavier than the *to* infinitive, so it makes some intuitive sense that its immediate repetition should be more strongly avoided than that of the infinitival variant.

The results on DCC-nesting are an instructive example of a spurious association and the fallibility of impressionistic intuitions. The 52 DCC-nested complements included 17 *to* infinitives and 35 gerundial complements, while the corresponding ratio for complements without such nesting was 554 to 384. Ignoring other variables, this translates to a marginal odds ratio of 0.34 and a corresponding logit of -1.09—a result which is statistically significant at $p < .001$ in the traditional Pearson's χ^2 test. Controlling for the other variables, however, the logit changes sign and becomes non-significant. The marginal association between DCC-nesting and *of -ing* appears to be due mainly to the Choice Principle: 36 or the 52 DCC-nested complements are [−Choice], and 32 of these involve the gerundial variant. In fact, the clausal complement of an instance of *afraid* premodified by *too* is frequently some feared consequence of the action described by the degree complement, as seen in (11):

(11)   Some bodies are still lying in the road as relatives are too afraid of getting killed themselves to retrieve their loved ones. (GB, 2016, *The Independent*)

The fact that the complement of *afraid* in such grammatical contexts is very often [−Choice] appears sufficient to account for the preponderance of gerundial complements, with the DCC itself exerting no additional impact.

4.4.2.3   Semantic Effects

Due to their interaction, [+Choice] and matrix negation are best analyzed together. The model estimates that the odds of a *to* infinitive in [+Choice] contexts with an unnegated higher clause are $e^{2.71} = 15$ times those in analogous [−Choice] contexts, and the effect is extremely statistically significant ($W = 9.54$; $p < .001$). Matrix negation is estimated to favor *to* infinitives at an odds ratio of $e^{2.03} = 7.6$ in [−Choice] contexts ($W = 5.36$; $p < .001$). However, this effect is greatly diminished in [+Choice] contexts, with an estimated odds ratio of only $e^{2.03-1.6} = 1.54$. The interaction is extremely statistically significant, with $G = 13.94$ and $p < .001$.

4.4.2.4   Other Effects

Taken collectively, outer and expanding-circle varieties appear to show a tendency for increased use of gerundial complements relative to inner-circle varieties. The odds ratio of $e^{0.57} = 1.76$ is statistically significant ($G = 9.74$; $p = 0.002$), but it is well worth remembering that statistical significance does not imply practical significance (Agresti 2015: 144). In particular, it makes little sense from a dialectological perspective to group together 11 different non-inner-circle varieties into a single category. However, since the research aims of the present chapter are not dialectological, such a simplified classification has served a purpose in helping reduce the risk that the effects of the main variables of interest might be confounded by potential syntactic peculiarities of non-inner-circle varieties. For the record, we also fit two alternative models to verify that no dramatic dialect differences involving inner-circle varieties were passing unnoticed. The first alternative model distinguished between all inner-circle varieties on the one hand and the single category of non-inner-circle varieties on the other. The second applied a ternary division between the two "American-like" varieties as a reference category that was contrasted to the two categories constituted by "British-like" inner-circle varieties and non-inner-circle varieties. Neither model reported dialect effects of significant magnitude between inner-circle varieties—the largest coefficient was always observed on the non-inner-circle category, and it was similar to the one reported by the main model seen above.

The estimated standard deviation of the verb-specific random intercepts is 0.59 logits. In other words, the average verb is estimated to favor one variant or another by an odds ratio of $e^{0.59} = 1.8$ compared to the overall mean. Expressed as an intracluster correlation, this implies that verb-specific idiosyncrasies account for an estimated $0.59^2/(\pi^2/3 + 0.59^2) = 9.4\%$ of the model's total explanatory power (Hosmer et al. 2013: 327). These numbers are considerably smaller than those reported for the Canadian and British data in Chapter 2. The random effect is statistically significant with a G-value of 5.48, one degree of freedom, and a p-value of 0.001,[7] but this is significantly due to the sample size. Note, especially, that in the Strathy dataset we had a larger random effect that nevertheless tested non-significant, doubtless due to the much smaller data size. Overall, verb-specific idiosyncrasies in the present dataset are fairly underwhelming, and the fixed-effect predictors do a good job of capturing the variability in the data.

4.5 CONCLUSION

The present chapter has presented nine potential explanatory variables of non-finite alternation under *afraid*, then proceeded to apply principles of purposeful sampling to construct a dataset facilitating reliable statistical inference on their effects, and finally estimated those effects using mixed-effects logistic regression. The variables of interest fall into three groups, representing processing complexity considerations, *horror aequi* factors, and semantic factors, respectively. Dialectal and lexical variation were treated as potential confounders.

Regarding processing complexity, the results are mostly in line with the predictions of previous univariate analyses, except where the Complexity Principle conflicts with another predictive principle. This is the case for passivization, whose most salient feature appears to be its tight association with patienthood rather than its processing complexity—passive complements, which are predominantly [−Choice], seem to attract the gerundial variant with the same intensity as active [−Choice] complements, despite their greater syntactic complexity.

The infinitive-favoring effect expected to follow from the increased syntactic complexity of negated complements appears to be canceled out by Hawkins' Minimize Domains Principle, which favors the gerundial complement in the relevant context. By contrast, the expected effects of insertions have been corroborated. Particularly noteworthy is the estimated effect differential between pre-complement insertions and complement-internal insertions, which requires further investigation. Likewise, the results lend robust support to the importance of the Extraction Principle, which conforms to expectation by strongly favoring the infinitival variant over the gerundial one.

Within the domain of *horror aequi*, a dramatic dispreference for sequences of two near-contiguous *of -ing* constructions was detected, but no analogous *horror aequi* of two quasi-adjacent *to* infinitives was found. It was speculated that the greater phonological complexity of the prepositional gerund may make its repetition more awkward than that of the infinitival variant. No evidence was found of *of -ing* being favored by *horror aequi* or production planning considerations in anticipation of degree complements licensed by *too*.

Within the semantic domain, the predictive validity of Choice Principle has been confirmed, with a dramatic infinitive-favoring effect reported by the statistical model. Another semantic factor showing promise in accounting for the variation is negation of the higher clause. *Not*-negation and *no*-negation were both found to favor *to* infinitives. Perhaps surprisingly, the model estimated a slightly stronger infinitive-favoring effect for the latter type of negation (cf. Rohdenburg 2015; Hagemeier 2006). However, the number of data points featuring *no*-negation was also much smaller than that of *not*-negation, so we stopped short of declaring its effect the stronger of the two.[8] From the standpoint of model parsimony, it seemed advantageous to treat the two types of negation as the same phenomenon. It is clear, however, that more research is needed to ascertain the relative effects of the two types of negation, and whether they truly are similar enough to warrant being lumped together.

As in Chapter 2, here too a strong interaction was found between the Choice Principle and matrix negation, and the emerging picture of the respective roles of the variables is substantively the same as in the BNC data. Given their respective effect sizes, it seems that [+Choice] is the primary infinitive-favoring factor between the two, while the effect of matrix negation appears to be largely limited to [−Choice] contexts.

No significant dialect differences were observed in complement choice between inner-circle varieties of English. A moderate statistically significant association was observed where non-inner-circle varieties, taken as a collective, displayed a slightly elevated rate of the gerundial variant relative to inner-circle varieties. Likewise, undramatic but statistically significant inter-verbal variation in complement choice was detected. We reserve judgment as to whether this implies that important grammatical factors are yet to be discovered, or whether a certain degree of lexical idiosyncrasy is always to be expected in syntactic variation phenomena, not just in the selectional preferences of complement-taking heads but also in the preferred morphosyntactic forms of lexical items themselves where more than one near-synonymous option is available.

The purposeful sampling technique is a powerful tool in the investigation of rare grammatical features and their impact on syntactic variation. The most challenging part of the sampling procedure employed in this chapter were the targeted searches for extractions, which occur in myriad syntactic contexts. In particular, it is doubtful whether our sample of extractions is fully representative of the true distribution of the subtypes of the phenomenon, i.e. whether the sample ratio of 174 complement extractions to only seven adjunct extractions is an accurate reflection of their relative frequencies in the population of extractions. Our search query may have better recall for the former type than for the latter. While we believe that the overall result—a strong infinitive-favoring effect for all extractions irrespective of subtype—is correct, our sample does not facilitate a reliable analysis of the possible differential effects of different extraction types. To begin to address this issue, the next chapter turns to an adjective whose natural co-occurrence rate with extractions is much higher than what we have observed with *afraid*.

NOTES

1. The authors owe the term *degree-complement construction* to Baltin (2006: 267). In their view, this designator is preferable to Huddleston and Pullum's (2002: 1262) "indirect complement", which may be incorrectly taken to imply that the degree complement is licensed by the adjective.

2. The authors wish to thank Mark Davies, who kindly granted permission for the first author to share concordances from the local copy of the corpus with the second author (p.c.).

3. All models were fit with 20 adaptive Gauss-Hermite quadrature points (Lesaffre and Spiessens 2001), employing optimization methods devised by Dennis and Schnabel (1983) and Schnabel et al (1985).

4. All *p*-values reported in this Section are based on the likelihood ratio test, which is known to be more reliable than the Wald test (Agresti 2013: 174–175).

5. Changing the sign of the log odds ratio yields an interpretation of the effect with the respective roles of the "success" and "failure" outcome reversed.

6. Though a vocative NP or a sentence boundary is not impossible after *of*, each of them can only occur under the severely limiting condition that the preposition be immediately followed by a gap left by interrogative, relativizing, or topicalizing extraction.

7. Since a standard deviation cannot be negative, the *p*-value of the likelihood-ratio test has been divided by two, following Agresti (2018: 278).

8. An earlier draft presented a more complex final model which included the interaction term and kept the two types of matrix negation separate. This model reported *not*-negation to have a much stronger infinitive-favoring effect than *no*-negation in [−Choice] contexts, but also a much stronger attenuation of this effect in [+Choice] contexts. However, the combination of an added interaction term and the division of the matrix negation data into two categories made the data so sparse that none of the results on *no*-negation were reliable.

APPENDIX—REGULAR EXPRESSIONS
USED IN CORPUS SEARCHES

General to Infinitive Search

(?i) (?<!too_rg)(?<!too_rg_rr@)afraid_\S+ ([^_]+_[rx]\S+
){0,2}(?#optional adverbs)(?#begin TO search)to_\S+ ([^_]+_[rx]\S+
){0,2}(?#optional adverbs)[^_]+_\S*v\S+(?#end TO search)

General of -ing Search

(?i) (?<!too_rg)(?<!too_rg_rr@)afraid_\S+ ([^_]+_[rx]\S+
){0,2}(?#optional adverbs)(?#begin TO search)to_\S+ ([^_]+_[rx]\S+
){0,2}(?#optional adverbs)[^_]+_\S*v\S+(?#end TO search)

Extractions, Active Voice

(?i) ([^_]+_\w\wq\S*(?#WH-word) (\S+_vb\S+ (?#NP search begin)([^_]+
_[adm](?!dq)\S+ ((kind|sort|type)_\S+ of_\S+ (a_\S+)?)?)?(?# Optional
determiner)([^_]+_(j|v[dv]g)\S+ ([^_]+_cc)?([^_]+_(j|v[dv]g)\S+
)?)?([^_]+_n\S+ ([^_]+_n\S+)?)?[^_]+_[ndp](?!px|dq)\S+(?# Head of
NP)(?# PP search begin)([^_]+_i\S+ ([^_]+_[adm]\S+)?(?# Optional
determiner)([^_]+(j|n|v[dv]g)\S+)?([^_]+_cc)?([^_]+_(j|n|v[dv]g)\S+
)?[^_]+_[nd]\S+){0,2}(?#NP search end)([^_]+_[rx]\S+)?(?#question
syntax, be VP)|\S+_v[mh]\S* (?#NP search begin)([^_]+_[adm](?!dq)\S+
((kind|sort|type)_\S+ of_\S+ (a_\S+)?)?)?(?# Optional determiner)([^_]+
_(j|v[dv]g)\S+ ([^_]+_cc)?([^_]+_(j|v[dv]g)\S+)?)?([^_]+_n\S+ ([^_]+
_n\S+)?)?[^_]+_[ndp](?!px|dq)\S+(?# Head of NP) (?# PP search
begin)([^_]+_i\S+ ([^_]+_[adm]\S+)?(?# Optional determiner)([^_]+
(j|n|v[dv]g)\S+)?([^_]+_cc)?([^_]+_(j|n|v[dv]g)\S+)?[^_]+_[nd]\S+
){0,2}(?#NP search end)([^_]+_[rx]\S+)?\S+_v\S+(?#question syntax,
modalized or perfective VP)|\S+_vd\S+ (\S+_xx)?(?#NP search
begin)([^_]+_[adm](?!dq)\S+ ((kind|sort|type)_\S+ of_\S+ (a_\S+
)?)?)?(?# Optional determiner)([^_]+_(j|v[dv]g)\S+ ([^_]+_cc)?([^_]+
_(j|v[dv]g)\S+)?)?([^_]+_n\S+ ([^_]+_n\S+)?)?[^_]+
[ndp](?!px|dq)\S+(?# Head of NP) (?# PP search begin)([^]+_i\S+
([^_]+_[adm]\S+)?(?# Optional determiner)([^_]+(j|n|v[dv]g)\S+
)?([^_]+_cc)?([^_]+_(j|n|v[dv]g)\S+)?[^_]+_[nd]\S+){0,2}(?#NP search
end)([^_]+_[rx]\S+)?\S+_v\S+(?#question syntax, lexical verb)) afraid_\S+

([^_]+_[rx]\S+){0,2}(?#optional adverbs)((?#begin TO search)to_\S+
([^_]+_[rx]\S+){0,2}(?#optional adverbs)[^_]+_\S*v\S+(?#end TO
search)|(?#begin OF ING search)of_\S+ ([^_]+_[rx]\S+){0,2}(?#optional
adverbs)(?!(any|every|no|some)thing_)(?!(any|every|no|some)thing_)[^_]+
ing_\S+(?#end OF ING search)) (\S+_nn\S+)?(?#optional
object)(\S+?_(i|rp)\S+)?(?#optional stranded preposition or
particle)[?]\S+(?#question search ends)|(?<!_v\w\w_jj)(?<!_v\w\w)([^_]+
_(nn|p|\w\wq|cs[an])\S+|that_\S+)(?#1-word NP antecedent) (that_\S+
|\S+?_\w\wq\S*)?(?#optional relative pronoun)(?#NP search begin)([^_]+
_[adm](?!dq)\S+ ((kind|sort|type)_\S+ of_\S+ (a_\S+)?)?)?)?(?# Optional
determiner)([^_]+_(j|v[dv]g)\S+ ([^_]+_cc)?([^_]+_(j|v[dv]g)\S+
)?)?)?([^_]+_n\S+ ([^_]+_n\S+)?)?)?[^_]+_[ndp](?!px|dq)\S+(?# Head of NP)
(?# PP search begin)([^_]+_i\S+ ([^_]+_[adm]\S+)?(?# Optional
determiner)([^_]+(j|n|v[dv]g)\S+)?([^_]+_cc)?([^_]+_(j|n|v[dv]g)\S+
)?[^_]+_[nd]\S+){0,2}(?#NP search end)(?#1 or 2-word subject)((([^_]+
_[rx]\S+)?\S+_v\S+ (?#lexical VP)|\S+_v\S+ ((?!too)[^_]+_[rx]\S+
){0,2}(\S+_v\S+)?(?#auxiliarized VP))afraid_\S+ ([^_]+_[rx]\S+
){0,2}(?#optional adverbs)((?#begin TO search)to_\S+ ([^_]+_[rx]\S+
){0,2}(?#optional adverbs)[^_]+_\S*v\S+(?#end TO search)|(?#begin OF
ING search)of_\S+ ([^_]+_[rx]\S+){0,2}(?#optional
adverbs)(?!(any|every|no|some)thing_)[^_]+ing_\S+(?#end OF ING
search)) (\S+?_(i|rp)\S+)?(?#optional stranded preposition or
particle)[!.?<]\S+(?#relativizer and comparativizer search ends))

Too-DCCs with an Embedded Complement of Afraid

(?i) too_\S+ afraid_\S+ ([^_]+_[rx]\S+)?(?#optional adverb)(to_\S+
([^_]+_[rx]\S+){0,2}(?#up to 2 optional adverbs)[^_]+_\S*v\S+|of_\S+
([^_]+_[rx]\S+){0,2}(?#up to 2 optional adverbs)\w+ing_\S+)
((?![.?!]_\S+ |[^_]+_(cs\S*|ccb\S*))(?# no intervening terminal
punctuation or subordinators allowed)\S+)*?to_\S+ ([^_]+_r\S+
){0,3}[^_]+_\S*v\S+

Extractions, Passive Voice

(?i) ([^_]+_\w\wq\S*(?#WH-word) (\S+_vb\S+ (?#NP search begin)([^_]+
_[adm](?!dq)\S+ ((kind|sort|type)_\S+ of_\S+ (a_\S+)?)?)?)?(?# Optional
determiner)([^_]+_(j|v[dv]g)\S+ ([^_]+_cc)?([^_]+_(j|v[dv]g)\S+
)?)?)?([^_]+_n\S+ ([^_]+_n\S+)?)?)?[^_]+_[ndp](?!px|dq)\S+(?# Head of

NP)(?# PP search begin)([^_]+_i\S+ ([^_]+_[adm]\S+)?(?# Optional
determiner)([^_]+(j|n|v[dv]g)\S+)?([^_]+_cc)?([^_]+_(j|n|v[dv]g)\S+
)?[^_]+_[nd]\S+){0,2}(?#NP search end)([^_]+_[rx]\S+)?(?#question
syntax, be VP)|\S+_v[mh]\S* (?#NP search begin)([^_]+_[adm](?!dq)\S+
((kind|sort|type)_\S+ of_\S+ (a_\S+)?)?)?)?(?# Optional determiner)([^_]+
_(j|v[dv]g)\S+ ([^_]+_cc)?([^_]+_(j|v[dv]g)\S+)?)?([^_]+_n\S+ ([^_]+
_n\S+)?)?[^_]+_[ndp](?!px|dq)\S+(?# Head of NP) (?# PP search
begin)([^_]+_i\S+ ([^_]+_[adm]\S+)?(?# Optional determiner)([^_]+
(j|n|v[dv]g)\S+)?([^_]+_cc)?([^_]+_(j|n|v[dv]g)\S+)?[^_]+_[nd]\S+
){0,2}(?#NP search end)([^_]+_[rx]\S+)?\S+_v\S+(?#question syntax,
modalized or perfective VP)|\S+_vd\S+ (\S+_xx)?(?#NP search
begin)([^_]+_[adm](?!dq)\S+ ((kind|sort|type)_\S+ of_\S+ (a_\S+
)?)?)?)?(?# Optional determiner)([^_]+_(j|v[dv]g)\S+ ([^_]+_cc)?([^_]+
_(j|v[dv]g)\S+)?)?([^_]+_n\S+ ([^_]+_n\S+)?)?[^_]+
[ndp](?!px|dq)\S+(?# Head of NP) (?# PP search begin)([^]+_i\S+
([^_]+_[adm]\S+)?(?# Optional determiner)([^_]+(j|n|v[dv]g)\S+
)?([^_]+_cc)?([^_]+_(j|n|v[dv]g)\S+)?[^_]+_[nd]\S+){0,2}(?#NP search
end)([^_]+_[rx]\S+)?\S+_v\S+(?#question syntax, lexical verb)) afraid_\S+
([^_]+_[rx]\S+){0,2}(?#optional adverbs)((?#begin TO search)to_\S+
([^_]+_[rx]\S+){0,2}(?#optional adverbs)(be|get)_\S+(?#end TO
search)|(?#begin OF ING search)of_\S+ ([^_]+_[rx]\S+){0,2}(?#optional
adverbs)(be|gett)ing_\S+(?#end OF ING search)) ([^_]+_r\S+
)?(?#optional adverb)\S+_v\wn\S* (?#past participle)(\S+_v\wg\S*
)?(?#optional present participle)([^_]+_(jj|v\wn|nn)\S*)?(?#optional
predicative or direct object)(\S+?_(i|rp)\S+)?(?#optional stranded
preposition or particle)[?]\S+(?#question search ends)|(?<!_v\w\w_jj
)(?<!_v\w\w)([^_]+_(nn|p|\w\wq|cs[an])\S*|that_\S+)(?#1-word NP
antecedent) (that_\S+ |\S+?_\w\wq\S*)?(?#optional relative
pronoun)(?#NP search begin)([^_]+_[adm](?!dq)\S+
((kind|sort|type)_\S+ of_\S+ (a_\S+)?)?)?)?(?# Optional determiner)([^_]+
_(j|v[dv]g)\S+ ([^_]+_cc)?([^_]+_(j|v[dv]g)\S+)?)?([^_]+_n\S+ ([^_]+
_n\S+)?)?[^_]+_[ndp](?!px|dq)\S+(?# Head of NP) (?# PP search
begin)([^_]+_i\S+ ([^_]+_[adm]\S+)?(?# Optional determiner)([^_]+
(j|n|v[dv]g)\S+)?([^_]+_cc)?([^_]+_(j|n|v[dv]g)\S+)?[^_]+_[nd]\S+
){0,2}(?#NP search end)(?#1 or 2-word subject)((([^_]+_[rx]\S+
)?\S+_v\S+ (?#lexical VP)|\S+_v\S+ ((?!too)[^_]+_[rx]\S+){0,2}(\S+_v\S+
)?(?#auxiliarized VP))afraid_\S+ ([^_]+_[rx]\S+){0,2}(?#optional
adverbs)((?#begin TO search)to_\S+ ([^_]+_[rx]\S+){0,2}(?#optional
adverbs)(be|get)_\S+(?#end TO search)|(?#begin OF ING search)of_\S+

([^_]+_[rx]\S+){0,2}(?#optional adverbs)(be|gett)ing_\S+_\S+(?#end OF ING search)) ([^_]+_r\S+)?(?#optional adverb)\S+_v\wn\S* (?#past participle)(\S+_v\wg\S*)?(?#optional present participle)([^_]+ _(jj|v\wn|nn)\S*)?(?#optional predicative or direct object)(\S+?_(i|rp)\S+)?(?#optional stranded preposition or particle)[!.?<]\S+(?#relativizer and comparativizer search ends))

General Supplementary Search for Passive Complements

(?i) (?<!too_rg)(?<!too_rg_rr@)afraid_\S+ ([^_]+_[rx]\S+){0,2}(?#optional adverbs)((?#begin TO search)to_\S+ ([^_]+_[rx]\S+){0,2}(?#optional adverbs)(be|get)_\S+(?#end TO search)|(?#begin OF ING search)of_\S+ ([^_]+_[rx]\S+){0,2}(?#optional adverbs)(be|gett)ing_\S+(?#end OF ING search)) ([^_]+_r\S+)?(?#optional adverb)\S+_v\wn\S*(?#past participle)

Immediately Preceding to Infinitive Search

(?i) (?<!_xx)to_\S+ \S+?_v(?!\wg)\S+ afraid_\S+ ([^_]+_[rx]\S+){0,2}(?#optional adverbs)((?#begin TO search)to_\S+ ([^_]+_[rx]\S+){0,2}(?#optional adverbs)[^_]+_\S*v\S+(?#end TO search)|(?#begin OF ING search)of_\S+ ([^_]+_[rx]\S+){0,2}(?#optional adverbs)(?!(any|every|no|some)thing_)[^_]+ing_\S+(?#end OF ING search))

Immediately Preceding of -ing Search

(?i) of_\S+ (?!(any|every|no|some)thing_)[^_]+ing_\S+ afraid_\S+ ([^_]+_[rx]\S+){0,2}(?#optional adverbs)((?#begin TO search)to_\S+ ([^_]+_[rx]\S+){0,2}(?#optional adverbs)[^_]+_\S*v\S+(?#end TO search)|(?#begin OF ING search)of_\S+ ([^_]+_[rx]\S+){0,2}(?#optional adverbs)(?!(any|every|no|some)thing_)[^_]+ing_\S+(?#end OF ING search))

Complement Negation Search

(?i) (?<!too_rg)(?<!too_rg_rr@)afraid_\S+ ((?#begin TO search)(to_\S+ not_xx|not_xx to_\S+) \S+?_v(?!\wg)\S+(?#end TO search)|(?#begin OF

ING search)of_\S+ (?#optional adverbs)not_xx
(?!(any|every|no|some)thing_)[^_]+ing_\S+(?#end OF ING search))

Pre-complement Insertion Search

(?i) (?<!too_rg)(?<!too_rg_rr@)afraid_\S+
((?!(never|of|to|sometimes_\S+ to_\S+ go|now_\S+ of_\S+ being)_)([^_]+
_[r,i]\S+|[^_]+_cs(?!t))){1,5}((?#begin TO search)to_\S+
\S+?_v(?!\wg)\S+(?#end TO search)|(?#begin OF ING search)of_\S+
(?#optional adverbs)(?!(any|every|no|some)thing_)[^_]+ing_\S+(?#end
OF ING search))

Intra-complement Insertion Search

(?i) (?<!too_rg)(?<!too_rg_rr@)afraid_\S+ ((?#begin TO search)to_\S+
((?!(never|of|to|even_\S+ attempt|even_\S+ dream_\S+ about|fully_\S+
reveal|openly_\S+ talk_\S+ about|publicly_\S+ criticize|even_\S+
ask|responsibly_\S+ create|permanently_\S+ losing|occasionally_\S+
putting|either_\S+ making(?#these eliminate already-existing
hits))_)([^_]+_[r,i]\S+|[^_]+_cs(?!t))){1,5}\S+?_v(?!\wg)\S+(?#end TO
search)|(?#begin OF ING search)of_\S+ ((?!(never|of|to|even_\S+
attempt|even_\S+ dream_\S+ about|fully_\S+ reveal|openly_\S+ talk_\S+
about|publicly_\S+ criticize|even_\S+ ask|responsibly_\S+
create|permanently_\S+ losing|occasionally_\S+ putting|either_\S+
making(?#these eliminate already-existing hits))_)([^_]+_[r,i]\S+|[^_]+
cs(?!t))){1,5}(?!(any|every|no|some)thing)[^_]+ing_\S+(?#end OF
ING search))

References

Agresti, Alan. 2013. *Categorical Data Analysis*, 3rd ed. Hoboken, NJ and
Chichester: Wiley; John Wiley [distributor].

Agresti, Alan. 2015. *Foundations of Linear and Generalized Linear Models*.
Hoboken, NJ: Wiley.

Agresti, Alan. 2018. *An Introduction to Categorical Data Analysis*, 3rd ed.
Hoboken, NJ: Wiley.

Baltin, Mark. 2006. Extraposition. In *The Blackwell Companion to Syntax*, vol.
II, ed. Martin Everaert and Henk van Riemsdijk, 237–271. Malden, MA:
Blackwell.

Biber, Douglas, Stig Johansson, Geoffrey Leech, Susan Conrad, and Edward Finegan. 1999. *Longman Grammar of Spoken and Written English*. London: Longman.

Bilder, Christopher R., and Thomas M. Loughin. 2015. *Analysis of Categorical Data with R*. Boca Raton: CRC Press.

Bolinger, Dwight. 1968. Entailment and the Meaning of Structures. *Glossa* 2: 119–127.

Dennis, John E., and Robert B. Schnabel. 1983. *Numerical Methods for Unconstrained Optimization and Nonlinear Equations*. Englewood Cliffs, NJ: Prentice-Hall.

Duan, Naihua, Dulal K. Bhaumik, Lawrence A. Palinkas, and Kimberly Hoagwood. 2015. Optimal Design and Purposeful Sampling: Complementary Methodologies for Implementation Research. *Administration and Policy in Mental Health and Mental Health Services Research* 42 (5) (09): 524–532.

Fox, John, and Sanford Weisberg. 2019. *An R Companion to Applied Regression*, 3rd ed. Thousand Oaks, CA: Sage.

Friedl, Jeffrey E. F. 2002. *Mastering Regular Expressions*, 2nd ed. Beijing and Sebastopol, CA: O'Reilly.

Garside, Roger, and Nicholas Smith. 1997. A Hybrid Grammatical Tagger: CLAWS4. In *Corpus Annotation: Linguistic Information from Computer Text Corpora*, ed. Roger Garside, Geoffrey Leech, and Tony McEnery, 102–121. London: Longman.

Givón, Talmy. 1990. *Syntax: A Functional-Typological Introduction*, vol. 2. Amsterdam: Benjamins.

Hagemeier, V. 2006. Der Einfluß struktureller/kognitiver Komplexität auf die Wahl alternativer Komplementstrukturen nach direktiven Ausdrücken im britischen und amerikanischen Englisch. MA thesis, University of Paderborn.

Hawkins, John A. 2004. *Efficiency and Complexity in Grammars*. Oxford: Oxford University Press.

Horn, Laurence R. 1978. Some Aspects of Negation. In *Universals of Human Language*, Vol. 4, ed. J. H. Greenberg, 127–210. Stanford: Stanford University Press.

Hosmer, David W., Stanley Lemeshow, and Rodney X. Sturdivant. 2013. *Applied logistic regression*, 3rd ed. Hoboken, NJ: Wiley.

Huddleston, Rodney D., and Geoffrey K. Pullum. 2002. *The Cambridge Grammar of the English Language*. 3rd printing 2010. Cambridge: Cambridge University Press.

Lavrakas, Paul J. 2008. *Encyclopedia of Survey Research Methods*. Thousand Oaks, CA: Sage.

Lesaffre, Emmanuel, and Bart Spiessens. 2001. On the Effect of the Number of Quadrature Points in a Logistic Random-Effects Model: An Example. *Journal of the Royal Statistical Society Series C* 50: 325–335. https://doi.org/10. 1111/1467-9876.00237.

Maier, Georg, Julia Davydova, and Peter Siemund. 2012. *The Amazing World of Englishes: A Practical Introduction.* Berlin: De Gruyter.

Quirk, Randolph, Sidney Greenbaum, Geoffrey Leech, and Jan Svartvik. 1985. *A Comprehensive Grammar of the English Language.* London: Longman.

Rohdenburg, Günter. 1996. Cognitive Complexity and Increased Grammatical Explicitness in English. *Cognitive Linguistics* 7 (2): 149–182.

Rohdenburg, Günter. 2003. Cognitive Complexity and *Horror Aequi* as Factors Determining the Use of Interrogative Clause Linkers in English. In *Determinants of Grammatical Variation in English*, ed. Günter Rohdenburg and Britta Mondorf, 205–249. Berlin: Mouton de Gruyter.

Rohdenburg, Günter. 2015. The Embedded Negation Constraint. In *Perspectives on Complementation: Structure, Variation and Boundaries*, ed. Mikko Höglund, Paul Rickman, Juhani Rudanko, and Jukka Havu. Houndmills, Basingstoke, Hampshire and New York, NY: Palgrave Macmillan.

Rohdenburg, Günter. 2016. Tracking Two Processing Principles with Respect to the Extraction of Elements out of Complement Clauses in English. *English Language and Linguistics* 20: 463–486.

Rudanko, Juhani. 2006. Watching English Grammar Change: A Case Study on Complement Selection in British and American English. *English Language and Linguistics* 10 (1): 31–48. https://doi.org/10.1017/s13606743060 01791.

Rudanko, Juhani. 2017. *Infinitives and Gerunds in Recent English: Studies on Non-finite Complements with Data from Large Corpora.* New York: Palgrave Macmillan.

Schnabel, Robert B., John E. Koontz, and Barry E. Weiss. 1985. A Modular System of Algorithms for Unconstrained Minimization. *ACM Transactions on Mathematical Software* 11: 419–440.

Shibatani, Masayoshi. 1985. Passive and Related Constructions. A Prototype Analysis. *Language* 61 (4): 821–848.

Tagliamonte, Sali A. 2006. *Analysing Sociolinguistic Variation.* Cambridge: Cambridge University Press.

Tagliamonte, Sali A. 2012. *Variationist Sociolinguistics: Change, Observation, Interpretation.* Malden, MA: Wiley-Blackwell.

Tottie, Gunnel. 1991. Lexical Diffusion in Syntactic Change: Frequency as a Determinant in the Development of Negation in English. In *Historical English Syntax*, ed. Dieter Kastovsky, 439–467. Berlin: Mouton de Gruyter.

Vosberg, Uwe. 2003a. The Role of Extractions and *Horror Aequi* in the Evolution of *-ing* Complements in Modern English. In *Determinants of Grammatical Variation in English*, ed. Günter Rohdenburg, and Britta Mondorf, 305–327. Berlin: Mouton de Gruyter.

Vosberg, Uwe. 2003b. Cognitive Complexity and the Establishment of -ing Constructions with Retrospective Verbs in Modern English. In *Insights into Late Modern English*, ed. Marina Dossena and Charles Jones, 197–220. Bern: Peter Lang.

Vosberg, Uwe. 2006. *Die Grosse Komplementverschiebung: Aussersemantische Einflüsse auf die Entwicklung Satzwertiger Ergänzungen im Neuenglischen.* Tübingen: Gunter Narr.

Wasow, Tom. 2002. *Postverbal Behavior.* Stanford, CA: CSLI Publications.

SOFTWARE USED

Anthony, Lawrence. 2018. AntConc (Version 3.5.7) [Computer Software]. Tokyo, Japan: Waseda University. Available from http://www.laurenceanth ony.net/software.

Bates, Douglas, Martin Maechler, Ben Bolker, and Steve Walker. 2015. Fitting Linear Mixed-Effects Models Using lme4. *Journal of Statistical Software* 67 (1): 1–48. https://doi.org/10.18637/jss.v067.i01.

Harrell, Frank E., with contributions from Charles Dupont and many others. 2018. Hmisc: Harrell Miscellaneous. R package version 4.1-1. https:// CRAN.R-project.org/package=Hmisc.

Nash, John C, and Ravi Varadhan. 2011. Unifying Optimization Algorithms to Aid Software System Users: Optimx for R. *Journal of Statistical Software* 43 (9): 1–14. http://www.jstatsoft.org/v43/i09/.

R Core Team. 2018. R: A Language and Environment for Statistical Computing. R Foundation for Statistical Computing, Vienna, Austria. https://www.R-pro ject.org/.

Wickham, Hadley. 2016. *ggplot2: Elegant Graphics for Data Analysis.* New York: Springer-Verlag.

CORPORA CONSULTED

Davies, Mark. 2013 *Corpus of News on the Web (NOW): 3+ Billion Words from 20 Countries, Updated Every Day.* Available online at https://corpus.byu.edu/ now/.

CHAPTER 5

Factors Bearing on Non-finite Complement Selection: A Case Study of *Accustomed* with Data from Hansard

Abstract The present study investigates non-finite complement choice of the adjective *accustomed* in Hansard transcripts of British Parliamentary debates. The objective is to inquire into the role of syntactic and semantic factors in the choice between infinitival and gerundial complement clauses with covert subjects at a time of significant variation between the two variants. The mid-twentieth century is identified as the time of greatest variation, and the twenty-year period from 1945 to 1964 is selected for closer quantitative analysis. A set of potentially explanatory factors, mainly syntactic and semantic, is investigated. Results of mixed-effects logistic regression suggest that extraction contexts, long insertions, and phonologically complex subordinate verbs exert a significant effect. In the domain of semantics, agentivity of the lower clause and negation of the higher clause prove consequential, and there are some indications of a significant interaction between the two factors.

Keywords Syntax of parliamentary language · Non-finite complementation · Variation · Multivariate analysis

© The Author(s) 2021
J. Ruohonen and J. Rudanko, *Infinitival vs Gerundial Complementation with Afraid, Accustomed, and Prone*,
https://doi.org/10.1007/978-3-030-56758-3_5

5.1   Introduction

This chapter investigates the nature and use of two non-finite complementation structures in Hansard, one particular text type of British English (see below). The patterns are initially illustrated by the sentences in (1a–b), from the Hansard Corpus.

(1a)   ... I am willing to trust Lord Diplock ... (1981)
(1b)   ... he is opposed to rearming Germany ... (1959)

In sentence (1a) the adjective *willing* selects a *to* infinitive complement, and in (1b) the adjective *opposed* selects what is here termed a *to -ing* complement. Both types of complement involve the word *to*, but as has been noted in the literature, the word *to* has two distinct syntactic functions in English: in sentence (1b) the word is a preposition, preceding a gerund, and in sentence (1a) it is what has often been called an infinitive marker, which is viewed as an auxiliary (or an Infl) for instance in Chomsky (1981). As noted by Warner (1993), the strongest piece of evidence for analyzing infinitival *to* as an auxiliary is supplied by VP Deletion, or post-auxiliary ellipsis, as Warner calls the rule. Consider the sentences in (2a–b), freely invented from (1a–b).

(2a)   I am willing to trust Lord Diplock but my colleague is not willing to.
(2b)   *He is opposed to rearming Germany, but his colleague is not opposed to.

Post-auxiliary ellipsis is typical of auxiliaries, as in *You won't do it, but I will*, and the well-formedness of (2a) is explained when infinitival *to* is accommodated under the Aux node (see Warner 1993: 64). By contrast, prepositional *to* is not an auxiliary, and therefore does not permit post-auxiliary ellipsis, explaining the contrast between (2a) and (2b). (For the auxiliary-like behavior of infinitival *to*, see also Radford 1997: 53.)[1]

There is no free variation between the two non-finite patterns in the case of *willing* and *opposed*, since both **willing to trusting Lord Diplock* and **opposed to rearm Germany* are ill formed. However, it has been pointed out that some other adjectives have shown variation and change between the two patterns. One such adjective is *accustomed*. For instance, consider the sentences in (3a–b).

(3a) People in the Services are accustomed to disregard political prejudices. (Hansard 1952)

(3b) We are accustomed to fertilising the land on which we grow crops ... (Hansard 1955)

In broad terms, it is the objective of this chapter to shed new light on the variation illustrated in (3a–b). It is assumed here that in each case the complement is sentential and has its own understood or covert subject. The postulation of a covert subject makes it possible to represent the argument structure of the lower sentence in a straightforward fashion. In the sentences in (3a–b), the matrix subjects are assigned their theta roles by *accustomed*, and the patterns therefore involve control, not movement. The sentences exhibit subject control, with the higher subjects controlling the understood subjects, represented with the symbol PRO in current work.

The variation of the type illustrated in (3a–b) is a natural choice for investigation because while both patterns are often illustrated side by side under the same sense of *accustomed* in major dictionaries,[2] many linguists also subscribe to Bolinger's Principle, which states that a "difference in syntactic form always spells a difference in meaning" (Bolinger 1968: 127). The adjective *accustomed* has been investigated before with respect to the variation between *to* infinitives and the *to* -*ing* pattern (see e.g. Kjellmer 1980; Vosberg 2006: 238–244; Rudanko 2006; Leech et al. 2009), and as regards diachronic change, the main lines in the evolution of its complement selection properties are by now well known. In the nineteenth century the *to* infinitive was much more frequent than the *to* -*ing* complement, but in accordance with one of the main features of the Great Complement Shift (see Rohdenburg 2006), *to* -*ing* complements gradually emerged, becoming the dominant pattern in the twentieth century, almost totally supplanting the infinitival pattern. Rudanko (2010) showed that in the TIME Corpus of American English the turning point can be located in the 1940s. By contrast, British English has been somewhat neglected in this line of work, and we wanted to investigate the adjective and factors impacting variation in its complement selection properties on the basis of the large Hansard Corpus, which was created as part of the SAMUELS project in the UK, and whose corpus architecture and search interface were provided by Mark Davies at Brigham Young University. To gain an overall picture, we began with tagged searches for the two types of complements of the adjective, with the search strings "accustomed to [vʔi*]" and "accustomed to [vʔg*]." While these search strings obviously do not retrieve all the relevant tokens, they do retrieve over 90%,[3] and they can be used to identify historical trends. The search strings yielded the information summed up in Fig. 5.1.

The pilot searches thus indicate that in the Hansard Corpus, the 1950s was a key decade of transition with respect to the two non-finite patterns, so we decided to investigate a 20-year period centered around that decade. To have maximum recall, we used the simple search string "accustomed." The selected subsection of the corpus, covering the period from 1945 to 1964, contains a total of 250 million words, and the number of tokens retrieved by our search string is 3,206. Among the tokens there are very frequent *to* NP complements, as in *We are accustomed to these examples of French gallantry* (1954), and there are also complements with overt subjects, as in *We are accustomed in this House to reports being pigeon-holed* (1955). While such types of complements deserve investigation, they are not the focus of the present study, and such tokens were identified by hand. After setting them aside, we were left with 1,405 tokens for investigation.

The main objective of the present investigation is to consider and to compare generalizations put forward as impacting variation between the two non-finite complements in the dataset of 1,405 tokens. Such generalizations are of different types. An important one pertaining to

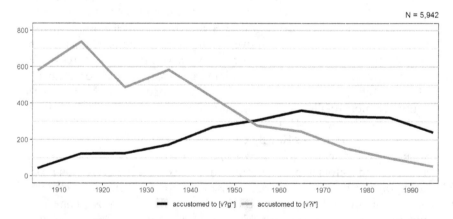

Fig. 5.1 Raw frequencies of the canonical forms of the two variants in the Hansard Corpus throughout the twentieth century

syntax is the Complexity Principle, which was first described by Günter Rohdenburg in the following terms:

> In the case of more or less explicit grammatical options the more explicit one(s) will tend to be favored in cognitively more complex environments. (Rohdenburg 1996: 151)

In the case of non-finite complements, the *to* infinitive has a higher cue validity as a sentential structure than gerunds (Rohdenburg 2016: 472), so the Complexity Principle predicts it to be favored in the presence of elements that increase processing load. One prominent manifestation of the phenomenon is the Extraction Principle. Pioneering work on that principle was undertaken by Günter Rohdenburg and Uwe Vosberg in the late 1990s and the 2000s. Vosberg (2003a) gave the following succinct definition:

> In the case of infinitival or gerundial complement options, the infinitive will tend to be favoured in environments where a complement of the subordinate clause is extracted (by topicalization, relativization, comparativization, or interrogation etc.) from its original position and crosses clause boundaries. (Vosberg 2003a: 308)

This definition limits extractions to the extraction of complements, but in later work the extraction of adjuncts has also been considered, including by Vosberg himself (Vosberg 2006, see also Rudanko 2006). For instance, consider (4a–b).

(4a)   ... those concerned can continue to sell the products which they have been accustomed to sell ... (1956)

(4b)   ... this was the most convenient centre for the shops, that it was where the people of Hull had been accustomed to do their shopping ... (1953)

In (4a) a complement has been extracted by Relativization, and it is easy to identify the extraction site. In (4b) it also seems possible to identify the extraction site (at the end of the extract quoted) and to posit that an adjunct has been extracted by Relativization. Both types of extraction are taken into account in this study, and the large number of tokens can be expected to yield new information on the roles played by the different types.

The second complexity factor analyzed in the present study is the insertion of material between the adjective and its complement. Insertions have been proposed by Vosberg (2003b: 210–211) as a factor favoring infinitival complements over gerundial ones, in accordance with the Complexity Principle. Following Wasow (2002: 38–41) and Berlage (2014), the present study investigates the effects of not only the presence but also the length and complexity of an insertion. Length is measured in words, as sanctioned by Szmrecsanyi (2004), while complexity is coded as a quaternary taxonomy contrasting complements immediately following the head with those preceded by matrix constituents, non-clausal insertion, or clausal insertion (whether finite or not). These three insertion types are illustrated in (5a–c):

(5a)   I do not intend to cross the Border further and to trespass on the territory of the North-East, however accustomed my forbears were to doing so some time ago ... (1963)

(5b)   In fact the police are quite accustomed in a moderate way to build up objections exactly on that issue[.] (1952)

(5c)   The position is that the Land Court has been accustomed, since it was set up in 1911, to deal with a large number of matters ... (1955)

In addition, due to the difference in syntactic function between auxiliary and prepositional *to*, it may be necessary to distinguish between pre-complement insertion—material intervening before the complement is chosen, i.e. before *to*—and within-complement insertion, which occurs when the insertion comes between *to* and the verbal head. However, the latter type seems rare with *accustomed*, and no instances of it were found in our dataset.

A semantic factor that has been identified in very recent work is the Choice Principle. It has been stated as follows:

In the case of infinitival and gerundial complement options at a time of considerable variation between the two patterns, the infinitive tends to be associated with [+Choice] contexts and the gerund with [−Choice] contexts. (Rudanko 2017: 20)

A [+Choice] context is then defined as a context where the (covert) subject of the sentential complement is an Agent. When the understood subject is not an Agent, the context is [−Choice].

The Choice Principle thus takes advantage of theta theory, an independent module of grammar, as the locus of a factor that may impact the choice of complement. The task of defining an Agent is far from trivial, and it is hardly possible to achieve complete unanimity among linguists regarding an optimal definition. Those working with the Choice Principle so far have used a cluster analysis, viewing agentivity as a cluster of features. In particular, they have given prominence to three features: to volitionality—the referent of the Agent is acting volitionally, or has "volitional involvement in the event or state," to recall a feature suggested by Dowty (1991: 572) for proto-agents, to control—the referent has at least some degree of control over the event or state, and to responsibility—the referent can be held responsible for the event or state. For instance, the lower predicate of sentence (3a), *disregard political prejudices*, as used in (3a), has an Agent as its subject—the people in question are conceptualized as acting volitionally, as being in control of their action and responsible for it. On the other hand, consider (6a–b):

(6a)   ... when driving up and down the country I have been accustomed to see such notices as "No one may stop here," ... (1951)

(6b)   ... we do not feel that he is treating the Committee entirely as it is accustomed to being treated. (1956)

Here, *see such notices as ...* and *being treated* exemplify predicates that are non-agentive. ("*Look at notices*" would normally be agentive, see Gruber 1967.) In them the referent of the subject is not volitionally involved in the event, nor in control of it, nor responsible for it.

There is an important correlation between [−Choice] and passivization. Subjects of passive clauses are overwhelmingly non-agents, given that the passive voice defocuses the agentive participant prototypically occupying subject position (Shibatani 1985). For this reason, the Choice Principle predicts a preponderance of gerunds in passive complements—a phenomenon that has been observed in the complementation of *afraid* (Rudanko 2015: 44; Sect. 2.3.2 of this volume). However, the passive is also a marked construction involving increased processing cost (Wanner

2009: 37), so the Complexity Principle predicts that the *to* infinitive should be preferred instead—a tendency that has likewise been claimed to hold true in some cases of non-finite alternation (Ball 1923: 100; Rohdenburg 2013: 124). It is therefore is advisable to include passivization in the multivariate analysis in order to help identify any independent influence that its syntactic markedness may exert upon complement choice under *accustomed*.

Another semantic factor sometimes linked to non-finite variation is negation of the higher clause, which was identified as a significant factor in Chapters 2 and 4 (see also Rickman and Rudanko 2018: 66–67). We include this variable in the analysis with the additional distinction that *no*-negation (Tottie 1991) is differentiated from *not*-negation.[4] In addition, features of the higher clause such as tense, modalization, subject type, and complexity of the subject NP have been associated with preference for the more explicit option in other complementation phenomena (Rohdenburg 2003; Ruohonen 2018). We make an effort to operationalize these factors and gain insight into their role in the alternation.

Potential effects of *horror aequi* (Rohdenburg 2003: 236–242), i.e. the putative avoidance of identical non-coordinated constructions in close succession (such as a hypothetical dispreference for *to* infinitives when the adjectival head is itself preceded by a *to* infinitive) is investigated to the extent permitted by the data. Finally, another factor proposed by Rohdenburg as a predictor of non-finite alternation under *accustomed* is the phonological complexity of the subordinate verb (2016: 474–475). This is measured in syllables and included in the analysis.

5.2   Description of the Dataset

The Hansard Corpus comprises 1.6 billion words of transcribed debates from the British Parliament, known collectively as the Hansard report. Annotated for part-of-speech and semantic information, the corpus spans two centuries from 1803 through 2005. Although the words "transcript" and "verbatim" appear in the description of the Hansard report (as of May 2019), the editing process that the original discourse undergoes on its way to the written record is substantial enough that the data may be more representative of written than of spoken language (Slembrouck 1992). This was the conclusion drawn by the compilers of the two (F)LOB corpora, both of which represent written discourse yet still include small sections from the Hansard report (Mollin 2007: 189). Due to these considerations, it seems appropriate to characterize the Hansard Corpus as a "speech-based corpus", i.e. a collection of texts which, while based on real-life speech events, are best described as reconstructions assisted by notes (Culpeper and Kytö 2010: 17).[5]

A total of 1,405 sentential complements of *accustomed* were identified in the 1945–1964 section of the corpus. Eligible exemplars have been illustrated in the previous section. Sentences (7a–b) exemplify tokens which were excluded from the sample due to not permitting the relevant variation.

(7a)   ... it is desirable that it should be confined to professional gentlemen who are accustomed to, and expert in, making decisions upon that point. (1948)

(7b)   Consequently, during that period, the whole Jewish population was being accustomed to act illegally, in revolt against the orders of the Administration of Palestine. (1946)

In (7a) *accustomed* is the first conjunct in coordination with *expert*, which does not license *to* infinitives, so only a gerund (or a NP) is possible here. In (7b) *accustomed* is a past participle of a transitive verb (the matrix predicate is passive), so the following *to* infinitive does not complement an adjective.

Figure 5.2 shows the marginal distributions of the competing variants by year. Despite significant random fluctuations in the number of

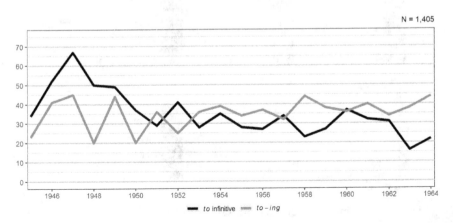

Fig. 5.2   Raw frequencies of the two variants by year, 1945–1964

yearly occurrences, an overall diachronic trend favoring *to -ing* is appreciable. Figure 5.3, in turn, cross-classifies complement type with every explanatory variable included in the ensuing multivariate analysis. The most striking univariate association in evidence is that between Extraction (both types) and the occurrence of the infinitival variant. Other conspicuous correlations include that between passive lower clauses and *to -ing*, the apparent attraction of polysyllabic lower verbs to the infinitive variant, and a dramatic bias for the gerundial variant in contexts where *accustomed* is preceded by a *to* infinitive (there were no data points with *accustomed* preceded by *to -ing*). There are regrettably few observations on this last phenomenon, which appears to be a manifestation of *horror aequi*. This scarcity of data will likely mean that no statistically significant results can be obtained on this variable, however, we will nonetheless enter prior *aequi* into the multivariate analysis in order to assess the extent to which

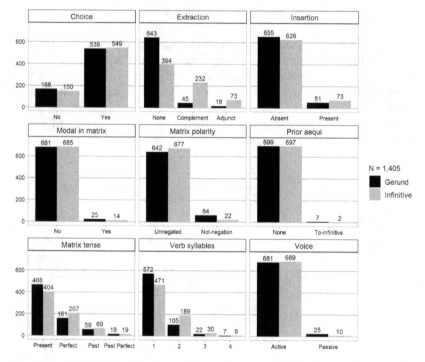

Fig. 5.3 Marginal distributions of the explanatory variables by complement type

the observed skew is attributable to other factors. The remainder of this chapter is devoted to such a multivariate analysis.

5.3   Multivariate Analysis—Model Selection

Treating *to* infinitives as the "success" outcome, we used the *lme4* package (Bates et al. 2015) in an R software environment to fit mixed-effects logistic regression models to the data.[6] An introduction to this statistical method has been given in Sect. 2.3.3. However, a number of features failed to occur in the dataset. As noted in Sect. 5.1 above, there were no complement-internal insertions. The same was true of gerundial *aequi*, i.e. not a single instance of *accustomed* complemented an immediately preceding *to -ing* structure.

We followed the stepwise model-building procedure detailed in Sect. 2.3.4. As per usual, our starting point was a full main-effects model containing those of the explanatory variables discussed in Sect. 5.1 that occurred in the dataset. The initial model was as follows:

Fixed Effects:

1. Extraction (ternary: None, Complement, Adjunct)
2. [±Choice] (dichotomous)
3. Voice (dichotomous)
4. Diachrony (quantitative: centered around the mean of 1954 and divided by ten)
5. Pre-complement insertion (quantitative: word count)
6. Type of pre-complement insertion (ternary: None, Phrasal, Clausal)
7. Matrix negation (ternary: None, *Not*-negation, *No*-negation)
8. Immediate prior *aequi* (dichotomous: None or *To* infinitive)
9. Phonological complexity of complement lemma (quantitative: syllable count centered around its median of 1)
10. Matrix modalization (ternary: None, Modal Verb, Modal Preterite)
11. Matrix tense (quaternary: Non-past, Preterite, Perfect, Preterite Perfect)
12. Matrix subject Head Type (ternary: Personal Pronoun, Other Pronoun, Noun)
13. Matrix distance (quantitative: number of words intervening between *accustomed* and the head of the matrix subject. This was intended as a surrogate for matrix-subject complexity).

Random Effects:

1. Subordinate Verb (nominal-scale with 292 categories)

As noted above, we followed the model-selection procedure detailed and illustrated in Chapter 2. A brief summary follows below. Our initial test for multicollinearity found just one issue, i.e. that insertion length in words was highly correlated with insertion type. This was to be expected because the two variables measured similar properties, i.e. the length and the syntactic complexity of an insertion, respectively. However, we concluded early on in the model-building process that only the word count of insertions mattered in complement choice, so only one variable from the collinear pair remained in the model.

Matrix modalization, matrix tense, type of matrix subject, and *no*-negation of the higher clause were all identified as immaterial in complement choice and subsequently dropped from the model. The same was not true for *not*-negation, which is discussed in more detail below. Once again, Voice proved inconsequential when [±Choice] was controlled for.

As for the variables that made it into the final model, we merged complement extraction and adjunct extraction into the same category— their respective coefficients of 2.56 and 2.32 were very similar in magnitude and both were statistically significant, so model parsimony dictated that the distinction be eliminated. The number of words intervening between the matrix subject and *accustomed* was not predictive when counted linearly. However, our routine check for non-linear effects detected an association between the gerundial variant and governors that followed the matrix subject immediately, without an intervening copula. We therefore replaced linear word counts with a dummy variable indicating whether such zero linkage was present. Likewise, a non-linear effect was found for the syllable count of the complement lemma. There, we found the most consequential difference to be between mono- and polysyllabic complement verbs, with the latter group favoring *to* infinitives. In contrast to these threshold phenomena, both diachrony and insertion length turned out to exert approximately linear effects on complement choice, the latter favoring *to* infinitives and the former, predictably, gerundial complements.

Aequi contexts where *accustomed* was immediately preceded by a *to* infinitive had a large gerund-favoring coefficient. It was not statistically significant—such contexts were found with only nine tokens. While it would have been justifiable to exclude these tokens from the analysis altogether (or to just drop the variable, barring confounding), we left this non-significant predictor in the model in order that it might stimulate future research into *horror aequi* in this or similar syntactic alternations. Finally, an interaction was once again detected between [+Choice] and *not*-negation of the higher clause. Thus, the final model stands as follows:

Fixed Effects:

1. Extraction (dichotomous)
2. Diachrony (quantitative: centered around the mean of 1954 and divided by ten)
3. Pre-complement insertion (quantitative: word count)
4. [±Choice] (dichotomous)
5. Matrix negation (dichotomous)
6. Polysyllabicity of complement verb (dichotomous)
7. Zero-linked governor (dichotomous)
8. Immediate prior *aequi* (dichotomous: none or *to* infinitive)

Interactions:

1. Choice × Matrix negation

Random Effects:

1. Subordinate verb (nominal-scale with 292 categories)

The next section interprets the output of this model.

5.4   MODEL INTERPRETATION

The chosen model has a simple classification accuracy of .767. Its concordance index equals .837, constituting "excellent" discrimination (Hosmer et al. 2013: 176). Figure 5.4 presents the estimated coefficients of the fixed effects in the final model, along with their 95% confidence intervals. The farther the confidence interval lies from zero, the more statistically significant the predictor, while an interval spanning zero implies $p > .05$.

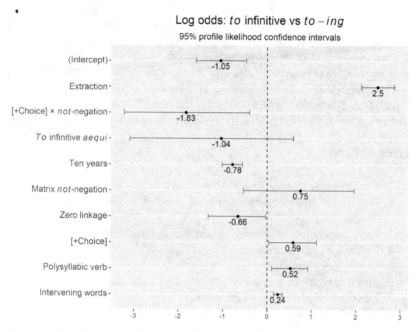

Fig. 5.4   Logit-scale coefficients of the variables included in the final model

The Intercept reports the estimated log odds of *to*-infinitival complements of *accustomed* at baseline, i.e. when all explanatory variables equal 0 (if quantitative) or their default category (if qualitative). This corresponds to a token occurring in 1954 without *not*-negation of the matrix clause, with at least one non-*to*-infinitival element intervening between the matrix subject and *accustomed*, and with nothing intervening between *accustomed* and its unagentive complement from which nothing has been extracted and whose main verb is monosyllabic. The odds of the *to* infinitive in such contexts are estimated at $e^{-1.05} = 0.35$, which translates to a probability of about .26.

The largest effect size is observed for Extraction, which is estimated to multiply the odds of a *to* infinitive by $e^{2.5} = 12$ relative to comparable contexts without extraction. The next-largest effect is exerted by insertions—with everything else equal, every word that intervenes between *accustomed* and its complement is estimated to multiply the odds of the *to* infinitive by $e^{0.24} = 1.27$. This translates to more than $e^{0.24*5} = 3.3$-fold *to*-infinitive odds for a five-word insertion, and $e^{0.24 \, *10} = 11.2$-fold odds for a ten-word insertion. The diachronic effect is entirely as expected, with the passage of time multiplying the odds of the gerundial variant by $e^{0.78} = 2.19$ in one decade, *ceteris paribus*.[7]

Due to their pairwise interaction, [+Choice] and *not*-negation of the matrix are best analyzed together. The model estimates that the odds of a *to* infinitive occurring in [+Choice] complements are $e^{0.59} = 1.8$ times the odds in analogous [−Choice] complements. The coefficients of *Not* and the interaction term suggest that, with other things equal, matrix *not*-negation multiplies the odds of an infinitival complement by $e^{0.75} = 2.12$ in [−Choice] contexts, but reduces them by a factor of $e^{0.75-1.83} = 0.34$ in [+Choice] contexts. It is worth noting that while fairly large, the coefficient of *Not* is not statistically significant ($W = .63$; $p = .23$). This is a reflection of the fact that there are only 14 negated matrices with a [−Choice] complement in the data, i.e. the estimate is based on such a small sample that it could be a coincidence. At any rate, *Not* must remain in the model since it is involved in a significant interaction (Agresti 2013: 210). As for the interaction itself, the high incidence of gerunds in agentive complements of negated matrices bears some resemblance to the observations made by Rudanko (2011: 180–185) on the same non-finite alternation in clausal complementation of the verb *commit*. In small samples of American and British English, he describes a preponderance of *to -ing* when the higher clause is negated. Recall that an interaction of [+Choice] and matrix negation was also found in the complementation of *afraid* in Chapter 2 (for British English) and Chapter 4. In those datasets, matrix negation simply failed to yield much of an additional infinitive-favoring effect if the context was [+Choice]. In the present dataset, the coefficients suggest something considerably more dramatic, i.e. that in [+Choice] contexts, matrix negation not only fails to reinforce but actually reverses the effect of agentivity.

It may be relevant here that due to the semantics of *commit*, its clausal complements are virtually always [+Choice]. They are thus semantically similar to [+Choice] complements in the present dataset. Broadly speaking, commitment and accustomedness both imply "propensity for" an action, so that declaring a referent committed or accustomed to an agentive activity conveys a positive epistemic bias in favor of the present or imminent actualization of that activity. The opposite is true for *afraid* (and other adjectives of fear) which, when predicated of the potential

agent of the situation, arguably convey a negative bias against its actualization. In both cases, negating the adjective reverses the bias. Accordingly,
not afraid with an agentive complement seems quite appropriate for
making threats, while the same is hardly true of the other two adjectives.
The opposite epistemic biases of propensity adjectives and fear adjectives
might be a factor in the different ways agentivity and matrix negation
interact in the complementation of the two types of adjective. Before
jumping to such conclusions, however, we should take note of the very
wide confidence intervals around the point estimates involving *Not*. The
estimates are imprecise due to the low number of negated matrices, especially in [−Choice] contexts. In fact, given these confidence intervals and
those reported in Sects. 2.3.5,[8] 2.4.4 and 4.4.2.3, it cannot be ruled
out that the true coefficients of matrix negation and its interaction with
[+Choice] might be the same for *afraid* and *accustomed*. At this stage,
the clearest difference between the two adjectives lies in the effect size
of [+Choice], not necessarily in the interaction between [+Choice] and
matrix negation.

Regarding the phonological complexity of the subordinate verb, the
model estimates that the odds of a *to* infinitive for polysyllabic lower
verbs are $e^{0.52} = 1.68$ times the odds for monosyllabic lower verbs, with
other things equal. This finding is broadly in line with that of Rohdenburg (2016: 747), who observes a similar tendency in a univariate analysis
limited to extraction contexts.

Zero linkage represents contexts where the higher subject is directly
linked to the adjective, as exemplified by (8a–c):

(8a)   We can not suddenly divert people accustomed to working
       inside a factory to such work as agriculture ... (1955)

(8b)   An efficient fishing industry must be based on ports accustomed
       to deal with a fishing fleet ... (1957)

(8c)   Where amenity problems occur, the technical inspectors sit with
       general inspectors accustomed to dealing with these matters ...
       (1958)

All such zero-linked instances of *accustomed* in our dataset postmodify
nouns, and they attract the gerundial variant at an odds ratio of $e^{0.66} =$
1.93 compared to analogous contexts where the adjective is the predicative complement of a verb. This invites speculation on whether modifiers,

being peripheral elements relative to heads, may be universally inclined toward less explicit clausal structures than heads, whenever two or more options of varying clausal explicitness are available. However, while statistically significant, the *p*-value is not overly impressive at .04, and it is by no means impossible that the observed effect might be an accident of random statistical noise.

Contexts where *accustomed* is immediately preceded by a *to* infinitive are estimated to favor the gerundial variant at an odds ratio of $e^{1.04} = 2.82$ with other things equal. Such a strong effect would certainly be in alignment with Rohdenburg's *horror aequi* principle—however, given the low number of *aequi* contexts in the dataset, the effect is not statistically significant. The *p*-value of .24 indicates that the finding may well be a fluke of this particular sample. It is worth pointing out that this factor had no effect on the complementation of *afraid* in the previous chapter, where it had been deliberately oversampled such that the non-significant result was based on a sample of over 60 observations rather than the present nine.

Table 5.1 lists the ten most infinitive-favoring verbs, while Table 5.2 presents the most gerund-favoring ones. The estimated standard deviation across lower verbs after controlling for the fixed effects is 0.77. In other words, the average verb is estimated to favor one or the other variant by an odds ratio of $e^{0.77} = 2.16$ relative to the overall mean. This is larger than most of the fixed effects, except for Extraction. Accordingly, the estimated intra-verb correlation suggests that $0.77^2/(\pi^2/3 + 0.77^2) = 15\%$ of the unexplained variance is attributable to verb-specific preferences

Table 5.1 The 10 most infinitive-favoring verbs in the dataset

Verb	Logit	Gerunds	Infinitives	n
expect	1.36	0	29	29
travel	1.08	0	6	6
drink	1.01	0	3	3
regard	0.82	2	12	14
trade	0.74	1	5	6
debate	0.73	0	3	3
rely	0.72	1	7	8
come	0.65	1	4	5
act	0.64	0	7	7
earn	0.61	0	3	3

Table 5.2   The 10 most gerund-favoring verbs in the dataset

Verb	Logit	Gerunds	Infinitives	n
discuss	−1.06	3	0	3
drive	−0.94	7	2	9
have	−0.82	40	17	57
get	−0.81	12	4	16
construe	−0.75	3	0	3
handle	−0.74	6	3	9
apply	−0.74	4	0	4
administer	−0.69	2	0	2
read	−0.65	10	4	14
keep	−0.64	8	2	10

(Hedeker and Gibbons 2006: 158), and the random effect is significant at $G = 15.55$, $df = 1$, $p < .001$. The most dramatic verb-specific effect is found for *expect*, which does not occur in *to -ing* even once despite its total of 29 appearances in the dataset.

It is difficult to discern any commonality within either group of verbs. Both groups seem to consist predominantly of verbs that are lexically agentive, though a few in both groups are construable either way. Both mono- and polysyllabic verbs are represented in both lists. It is also worth pointing out that *discuss* and *debate* show the opposite biases despite being close in meaning. We suspect that there is probably no rhyme or reason to this chaos, and that a significant proportion of these apparent lexical idiosyncrasies will melt away in future analyses where idiolect is controlled.

5.5   CONCLUSION

The present chapter has employed multivariate analysis to investigate conditioning factors of the free variation between two non-finite sentential complements of the adjective *accustomed* in transcripts of British parliamentary debates at a time of considerable variation between the two patterns, i.e. the mid-twentieth century. The results lend robust support to the Complexity Principle, especially evident in the large and statistically significant effect of Extraction, which exhibits a striking preference for the infinitival, i.e. more explicitly clausal, constructional variant. Unlike in Chapter 4, where adjunct extractions were too few to be reliably analyzed, the present dataset contained 91 such extractions, and the results indicate that it makes little difference in the complementation of *accustomed* whether the extracted constituent is a complement or an adjunct.

A somewhat less dramatic, though no less interesting, complexity effect has been observed for Insertions, whose predilection for the *to* infinitive displays an approximately linear relationship to the insertion's length in words. A third, if again less dramatic, complexity effect is the tendency for polysyllabic complement verbs to favor the more explicitly clausal option, consistent with preliminary indications in previous literature.

By contrast, the modality and tense of the matrix clause and the length and type of the matrix subject appear to have no bearing on complement choice with *accustomed*. The same can also be said for passivization of the lower clause, whose potential for a complexity effect appears to be overruled by semantics—more specifically by the Choice Principle. The Choice Principle correctly predicts that gerunds are favored in unagentive clauses, i.e. in the semantic type of clause that passives prototypically encode. This is indeed the case in our dataset, where [−Choice] complements display a moderate, statistically significant preference for the gerundial option regardless of voice.

Another consequential factor in the alternation appears to be the negation of the higher clause. An intriguing interaction was detected between [+Choice] and matrix negation, whereby negated matrices favored the infinitival option in [−Choice] contexts, but the effect changed direction with [+Choice] complements. However, despite the large overall size of the dataset, negated matrices with [−Choice] complements are too few to warrant confident statements about their behavior. Suffice it to say that negation of the higher clause appears to play a role in clausal complementation, evidently interacting with the agentivity of the complement, and both phenomena deserve further investigation. The same goes for *horror aequi*, for whose reliable estimation the present dataset is lamentably insufficient, but whose statistically non-significant effect is directionally concordant with what earlier work predicts.

Perhaps the most difficult effect to explain is the observed tendency for zero-linked predicates to favor *to -ing*. We have speculated that the noun postmodifier status of such predicates, obviously a lower rank in the syntactic hierarchy than their prototypical role as the head of a predicative complement, may have something to do with the increased incidence of the less explicitly clausal option. Such a tendency would certainly accord with Hopper and Thompson's suggestion that more nominalized clauses are associated with backgrounding (1980: 285). Though this hypothesis has some intuitive appeal, some skepticism is in order—given the ease with which "statistical significance can be obtained from pure noise" (McShane et al. 2019: 2), it seems prudent to remain agnostic about the reality of this phenomenon until it has been replicated in other datasets.

Finally, there remains a fair amount of unexplained variance that is attributable to individual verbs, some of which exhibit striking biases for one variant or another. We believe that a considerable proportion of these "lexical effects" is due to confounding by one or more important variables that could not be included in the present model, either because the information was not available (transcriber identity) or because they have not yet been identified. Much work remains to be done before all the important predictors of non-finite complement selection are fully understood.

NOTES

1. However, the present authors do not regard infinitival *to* as an affix, because it can easily be separated from the following verb, as in *willing to not trust Lord Diplock*.

2. At the same time, the second edition of the *OED* from 1989 only illustrated *to* infinitives of the adjective, under the verb *accustom*. The third edition, from 2011, illustrates both types.

3. This search found 1,280 of the 1,405 observations retrieved by the more general search conducted for the statistical analysis in Sect. 5.2.

4. We also wanted to study the effects of complement negation, which has been shown to influence complement choice in both finite and non-finite alternation phenomena (Ruohonen 2018; Rohdenburg 2015). However, this had to be abandoned because the search results included only one negated complement. The lone token was excluded from the quantitative analysis.

5. Mollin (2007) gives an example relating to non-finite complementation of how the text of Hansard may differ from the original speeches by observing that the verb *help*, when followed by an infinitive without *to* in her original material, was regularly replaced with the *help to* infinitive construction in Hansard.

6. Each model was fit by the *nlm* optimization algorithm in the *optimx* package (Nash and Varadhan 2011), using optimization ideas of Dennis and Schnabel (1983) and Schnabel et al. (1985). The number of adaptive Gaussian quadrature points used in each fitting process was 20.

7. Changing the sign of the coefficient yields an analogous interpretation of the effect in which the respective roles of the success and failure outcome are reversed.
8. The point estimates of 1.34 and −1.03 were non-significant in the Strathy dataset, implying confidence intervals extending to both sides of zero.

REFERENCES

Agresti, Alan. 2013. *Categorical Data Analysis*, 3rd ed. Hoboken, NJ and Chichester: Wiley; John Wiley [distributor].

Ball, F.K. 1923. *Constructive English: A Handbook of Speaking and Writing*. Boston: Ginn & Co.

Bates, Douglas, Martin Maechler, Ben Bolker, and Scott Walker. 2015. Fitting Linear Mixed-Effects Models Using lme4. *Journal of Statistical Software* 67 (1): 1–48. https://doi.org/10.18637/jss.v067.i01.

Berlage, Eva. 2014. *Noun Phrase Complexity in English*. Cambridge: Cambridge University Press.

Bolinger, Dwight. 1968. Entailment and the Meaning of Structures. *Glossa* 2: 119–127.

Chomsky, Noam. 1981. *Lectures on Government and Binding*. Dordrecht: Foris.

Culpeper, Jonathan, and Merja Kytö. 2010. *Early Modern English Dialogues: Spoken Interaction as Writing*. Cambridge and New York: Cambridge University Press.

Dennis Jr., John E., and Robert B. Schnabel. 1983. *Numerical Methods for Unconstrained Optimization and Nonlinear Equations*. Englewood Cliffs, NJ: Prentice-Hall.

Dowty, David. 1991. Thematic Proto-Roles and Argument Selection. *Language* 67 (3): 547–619.

Gruber, Jeffrey. 1967. Look and See. *Language* 43 (4): 937–947.

Hedeker, Donald, and Robert D. Gibbons. 2006. *Longitudinal Data Analysis*. Hoboken, NJ: Wiley-Interscience.

Hopper, Paul J., and Sandra Thompson. 1980. Transitivity in Grammar and Discourse. *Language* 56 (2): 251–299.

Hosmer, David W., Stanley Lemeshow, and Rodney X. Sturdivant. 2013. *Applied Logistic Regression*, 3rd ed. Hoboken, NJ: Wiley.

Kjellmer, Göran. 1980. Accustomed to Swim; Accustomed to Swimming: On Verbal Forms After TO. In *ALVAR: A Linguistically Varied Assortment of Readings. Studies Presented to Alvar Ellegård on the Occasion of his 60th Birthday*, ed. Jens Allwood and Magnus Ljung, 175–190. Stockholm: Almqvist and Wiksell.

Leech, Geoffrey N., Marianne Hundt, Christian Mair, and Nicholas Smith. 2009. *Change in Contemporary English: A Grammatical Study*. Cambridge: Cambridge University Press.

McShane, Blakeley B., David Gal, Andrew Gelman, Christian Robert, and Jennifer Tackett. 2019. Abandon Statistical Significance. *The American Statistician* 73 (1): 235–245.

Mollin, Sandra. 2007. The Hansard Hazard. Gauging the Accuracy of British Parliamentary Transcripts. *Corpora* 2 (2): 187–210.

Nash, John C., and Ravi Varadhan. 2011. Unifying Optimization Algorithms to Aid Software System Users: Optimx for R. *Journal of Statistical Software* 43 (9): 1–14. http://www.jstatsoft.org/v43/i09/.

R Core Team. 2018. R: A Language and Environment for Statistical Computing. R Foundation for Statistical Computing, Vienna, Austria. https://www.R-project.org/.

Radford, Andrew. 1997. *Syntactic Theory and the Structure of English*. Cambridge: Cambridge University Press.

Rickman, Paul, and Juhani Rudanko. 2018. *Corpus-Based Studies on Nonfinite Complements in Recent English*. Houndmills, Basingstoke: Palgrave Macmillan.

Rohdenburg, Günter. 1996. Cognitive Complexity and Increased Grammatical Explicitness in English. *Cognitive Linguistics* 7 (2): 149–182.

Rohdenburg, Günter. 2003. Cognitive Complexity and *Horror Aequi* as Factors Determining the Use of Interrogative Clause Linkers in English. In *Determinants of Grammatical Variation in English*, ed. Günter Rohdenburg and Britta Mondorf, 205–249. Berlin: Mouton de Gruyter.

Rohdenburg, Günter. 2006. The Role of Functional Constraints in the Evolution of the English Complementation System. In *Syntax, Style and Grammatical Patterns*, ed. Christina Dalton-Puffer et al., 143–166. Bern: Peter Lang.

Rohdenburg, Günter. 2013. The Construction *Cannot Help -ing* and Its Rivals in Modern English. In *Corpus Perspectives on Patterns of Lexis*, ed. Hasselgård et al., 113–132. Amsterdam and Philadelphia: John Benjamins.

Rohdenburg, Günter. 2015. The Embedded Negation Constraint. In *Perspectives on Complementation: Structure, Variation and Boundaries*, ed. Höglund et al., 101–127. Houndmills, Basingstoke, Hampshire and New York, NY: Palgrave Macmillan.

Rohdenburg, Günter. 2016. Testing Two Processing Principles with Respect to the Extraction of Elements out of Complement Clauses in English. *English Language and Linguistics* 20: 463–486.

Rudanko, Juhani. 2006. Watching English Grammar Change. *English Language and Linguistics* 10: 31–48.

Rudanko, Juhani. 2010. Exploring Grammatical Variation and Change: A Case Study of Complementation in American English over Three Decades. *Journal of English Linguistics* 38 (1): 4–24.

Rudanko, Juhani. 2011. *Changes in Complementation in British and American English: Corpus-Based Studies on Non-Finite Complements in Recent English.* Houndmills, Basingstoke, Hampshire: Palgrave Macmillan.

Rudanko, Juhani. 2015. *Linking Form and Meaning: Studies on Selected Control Patterns in Recent English.* London: Palgrave Macmillan.

Rudanko, Juhani. 2017. *Infinitives and Gerunds in Recent English: Studies on Non-Finite Complements with Data from Large Corpora.* New York: Palgrave Macmillan US.

Ruohonen, Juho. 2018. 21th-Century Trends in BrE Mandative Constructions—A Statistical Multivariate Analysis. Paper presented at the Fifth International Conference of the International Society for the Linguistics of English (ISLE), London, UK.

Schnabel, Robert B., John E. Koontz, and Barry E. Weiss. 1985. A Modular System of Algorithms for Unconstrained Minimization. *ACM Transactions on Mathematical Software* 11: 419–440.

Shibatani, Masayoshi. 1985. Passive and Related Constructions. A Prototype Analysis. *Language* 61 (4): 821–848.

Slembrouck, Stef. 1992. The Parliamentary Hansard 'Verbatim' Report: The Written Construction of Spoken Discourse. *Language and Literature* 1 (2): 101–119.

Tottie, Gunnel. 1991. Lexical Diffusion in Syntactic Change: Frequency as a Determinant in the Development of Negation in English. In *Historical English Syntax*, ed. Dieter Kastovsky. Berlin: Mouton de Gruyter.

Vosberg, Uwe. 2003a. The Role of Extractions and *Horror Aequi* in the Evolution of *-ing* Complements in Modern English. In *Determinants of Grammatical Variation in English*, ed. Günter Rohdenburg, and Britta Mondorf, 305–327. Berlin: Mouton de Gruyter.

Vosberg, Uwe. 2003b. Cognitive Complexity and the Establishment of *-ing* Constructions with Retrospective Verbs in Modern English. In *Insights into Late Modern English*, ed. Marina Dossena and Charles Jones, 197–220. Bern: Peter Lang.

Vosberg, Uwe. 2006. *Die Grosse Komplementverschiebung.* Tübingen: Narr.

Wanner, Anja. 2009. *Deconstructing the English Passive.* Berlin and New York: Mouton de Gruyter.

Warner, Anthony. 1993. *English Auxiliaries: Structure and History.* Cambridge: Cambridge University Press.

Wasow, Thomas. 2002. *Postverbal Behavior.* Stanford, CA: CSLI Publications.

CORPORA CONSULTED

Davies, Mark. 2015. Hansard Corpus. Part of the SAMUELS project. Available online at https://www.hansard-corpus.org/.

CHAPTER 6

Factors Bearing on Infinitival and Gerundial Complements of the Adjective *Prone* in Current American and British English

Abstract The present study investigates non-finite complement choice of the characterizing modal adjective *prone* in British and American English in the relevant subsections of the NOW Corpus. The objective is to inquire into the role of syntactic and semantic factors in the choice between infinitival and gerundial complement clauses with covert subjects at a time of significant variation between the two variants, i.e. the mid-2010s. A set of potentially explanatory factors, mainly syntactic and semantic, are investigated. The strongest predictors of complement choice appear to be semantic. Stativity, future reference, and repeatability of the complement situation are all highly significant predictors. The animacy, agentivity, and countability of the lower subject are likewise consequential. Within the syntactic domain, extraction contexts and propositional (rather than VP-internal) scope of the adjective are highly significant predictors. Lastly, the phonological complexity of the complement verb is also found to play a role in variant selection. British English turns out be more favorable to the gerundial complement than its colonial offshoot, and the difference is highly statistically significant.

Keywords Syntax · Non-finite complementation · Variation · Multivariate analysis

J. Ruohonen and J. Rudanko, *Infinitival vs Gerundial Complementation with Afraid, Accustomed, and Prone*,
https://doi.org/10.1007/978-3-030-56758-3_6

6.1   INTRODUCTION

Consider the sentences in (1a–c), from the American and British sections of the NOW Corpus.

(1a)   McPhee is more prone to stress the agony of composition... (US, 2017, *The Weekly Standard*)

(1b)   Frankie seems prone to trying almost anything. (US, 2017, *The Ringer*)

(1c)   UKIP's vote share is prone to be squeezed, benefiting the Conservatives. (GB, 2015, *The Independent*)

Sentences (1a–c) are similar in that in all three the adjective *prone* selects a sentential complement. In (1a) and (1c) the complement is a *to* infinitive, and in (1b) it is a *to -ing* clause, to use the terminology of Chapter 5. The sentences in (1a–c) show that the adjective *prone* selects both types of complements in current English with ease. In broad terms, the purpose of this chapter is to investigate the variation between these two sentential patterns in the case of the matrix adjective *prone* and the factors that may have a bearing on that variation.

An important similarity between the complements in (1a–c) is that the adjective *prone* assigns a theta role to the subject of the higher sentence. In accordance with the assumption made in Chapter 1 of this volume that infinitival and *to -ing* complements involve understood subjects, the understood subjects are then represented with the symbol PRO. Both patterns involve the word *to*, but in accordance with the analysis of Sect. 5.1, it is taken for granted that the type of *to* in (1a) and (1c) is syntactically different from the *to* of sentence (1b). The former may be called infinitival *to*, and the latter is prepositional *to*. Since the *to* of (1b) is a preposition, it is helpful to use the traditional notion of a nominal clause in the analysis of (1b), with the *to -ing* complement involving such a clause, that is, a sentence dominated by a NP. It may be added that the adjective *prone* also commonly selects complements consisting of the preposition *to* and a non-sentential NP, as in ... *people prone to depression are just likelier to end up in those jobs* (AmE segment of the NOW Corpus).

Accepting the assumptions outlined in the paragraph above, the syntactic structures of (1a–c) may be represented in their essentials as in (1a′–c).

(1a′)   [[McPhee]$_{NP}$ is [[more prone]$_{AdjP}$ [[PRO]$_{NP}$ [to]$_{Aux}$ [stress the agony of composition]$_{VP}$]$_{S2}$]$_{AdjP}$]$_{S1}$

(1b′)   [[Frankie]$_{NP}$ seems [[prone]$_{Adj}$ [[to]$_{Prep}$ [[[PRO]$_{NP}$ [trying almost anything]$_{VP}$]$_{S2}$]$_{NP}$]$_{PP}$]$_{AdjP}$]$_{S1}$

(1c′)   [[Ukip's vote share]$_{NP}$ is [[prone]$_{Adj}$ [[PRO]$_{NP}$ [to]$_{Aux}$ [be squeezed]$_{VP}$]$_{S2}$]$_{AdjP}$]$_{S1}$

The representations in (1a′–c′) are analogous to the representations of the same types of complements selected by the adjective *accustomed*, as outlined in Chapter 5.

The oldest attested use of *prone* listed in the *OED* is dated around 1385 and now marked obsolete. It is glossed as "[h]aving a downward aspect or direction; having a downward or descending inclination or slope; (also) steeply or vertically descending, headlong." The use of *prone* with nominal and sentential *to* complements appears to be metaphorically linked to the positional sense, first attested a decade after the positional sense and glossed in the *OED* as sense I 1: "[h]aving an inclination or tendency to something; (naturally) disposed, inclined, or liable". This co-existence of an older physical sense with an extended one relating to "natural" tendencies suggests that with sentential complements, the primary, prototypical use of *prone* is as a characterizing adjective signaling dynamic modality. Characterizing predicates are a subtype of generics, and they describe "essential", i.e. regular or predictable behaviors of entities (Krifka et al. 1995: 2–8), while dynamic modality is a subtype of modality that concerns conditioning factors that are internal to the relevant entity (Palmer 2001: 9).[1] Characterizing genericity and modality vary independently—*she writes novels* is characterizing and non-modal, while *she wants to fire Kim* is modal and non-characterizing—but *prone* prototypically conveys both dynamicity and characterization as elements of its lexical meaning. However, as is suggested by the inclusion of 'liable' among the *OED*'s glosses, *prone* can also be used to signal epistemic modality. This is exemplified by (1c), which occurred in an election forecasting context. Here, the use of *prone* is clearly not characterizing, since a party's vote share in a particular election can hardly be attributed "regular" behaviors. Rather, the author is making a prediction on how a particular situation might play out. However, such epistemic uses seem to be subordinate. This is supported by the fact that *prone* is incompatible with some of the syntactic behaviors of its full-time epistemic colleagues. For example, it disallows extraposition, and it is hardly acceptable with dummy subjects:

(2a)   It is likely that Ukip's vote share will be squeezed.
(2b)   *It is prone that Ukip's vote share will be squeezed.
(2c)   There is likely to be a series of very small controlled explosions in the area. (GB, 2015, *The Press, York*)
(2d)   ??There is prone to be a series of very small controlled explosions in the area.

The *OED* supplies numerous illustrations of *to* infinitives. Virtually all of them come under the subdivisions of (a) and (b) of the aforementioned modal sense, listing uses that relate to negative and positive complement situations, respectively:

(3a)   All are by nature prone to err. (*OED*, 1883, B. Jowett tr. *Hist Peloponnesian War*).
(3b)   American writers were prone to make much of the compact signed in the cabin of the Mayflower. (1883, *Cent. Mag.*)

Examples of *to -ing* complements are far fewer in the *OED*, and among the illustrations from the last three centuries, there is only one that may be viewed as being of that type. It is given in (4).

(4)   His stage manner is wooden and he's much too prone to sobbing to make the emotional points. (*OED*, 2004, *Opera Now*)

The *-ing* form in sentence (4) appears to be ambiguous between a gerundial and a nominal interpretation, but even if it is admitted as a verbal complement, as the present authors are inclined to do, the number of *to* infinitival examples is much larger, creating the impression that the *to* infinitive is much more frequent than the gerundial complement with the adjective. In the same spirit, Poutsma (MS), writing in the 1930s, only cites one example of what is here called the *to -ing* complement, using the label "to + ger.," which is *She was a woman prone to quarrelling* (from Trollope), but numerous examples of *to* infinitives. The main focus of this chapter is not on the frequencies of the two patterns in current English, but it may still yield information on whether the gerundial pattern has established itself in English with *prone*, in the spirit of the Great Complement Shift.

The broader point to emerge from the *OED*, and also from Poutsma (MS), is that both non-finite patterns are treated under the same sense of the adjective. This circumstance makes the adjective worth exploring from the point of view of Bolinger's Principle. If the sense of the adjective can be taken to be fairly constant, it is then possible to focus on the nature and impact of various other factors that may influence the choice of complement with *prone*.

As in previous chapters, one major focus of the statistical analysis consists in testing the validity of the Extraction Principle, the Choice Principle, and a number of other syntacto-semantic factors that have been put forth in previous literature. In addition, contextual variation in the salience of the different components, or subsenses, of the adjective's meaning raises questions about their possible differential complementation preferences. Lastly, in contrast to the adjectives treated in previous chapters, *prone* allows inanimate and abstract subjects. The potential consequences that the variable animacy of the subject may have for complement selection will also be investigated.

6.2   DATA

As in Chapters 3 and 4 above, our data comes from of a local copy of the NOW Corpus that is part-of-speech tagged with the Claws 7 tagset and extends diachronically from January 2010 to the end of October 2017.[2] Focusing on the British and American sections, we searched for any and all sentences with an instance of *prone* that could conceivably be followed by a complement clause. Our regular expression is seen below. It matches *prone*, followed by *to* at any distance within the same sentence, followed by anything that could conceivably be an infinitive or gerund within the same sentence:

```
(?i)\sprone_\S+\s((?![.!?]_)\S+\s)*?to_\S+\s((?![.!?]_)\S+\s)*?\S+_v\w[^dnzrm]\S*
```

In order to maximize data quality while keeping the workload within manageable bounds, we first sorted the concordance lines from both varieties by date, then systematically included eligible tokens, starting from the newest, until the total for that variety reached 750. This resulted in the largest dataset analyzed in the present volume, i.e. a database of 1,500 sentential complements of *prone*, evenly split between the two varieties. The AmE data spans a period from February 2015 through October 2017. The British section of the corpus is diachronically sparser than the American, so the BrE tokens represent a slightly longer time span, from April 2014 to October 2017.

We excluded obvious duplicates, i.e. cases where the same sentence is first used in the main text and then re-used verbatim as an image caption within the same article (or the other way around). Other rejects include coordinations with a pure noun, contexts where only a nominal interpretation of the formally ambiguous word makes sense, and the doubl-*ing* constraint (Ross 1972). Examples are shown below:

(5a)   DeLaria, a Broadway veteran, was prone to belting and theatricality. (US, 2017, *Minneapolis Star Tribune*)

(5b)   Karst is a porous, unstable type of rock prone to collapse and sinkholes. (US, 2017, *Lynchburg News and Advance*)

(5c)   Remember how H3N2 is especially prone to drift? (US, 2015, *WIRED*)

(5d)   No Wall Street bosses ever were jailed, so they're prone to keep speculating with your money... (US, 2017, *Common Dreams*)

Since coordinates usually belong to the same syntactic category (Huddleston and Pullum 2002: 1290), it seems safest to regard (5a) as a coordination of two NPs, each consisting of a noun. Infinitival *to* does not take nominal complements, and the coordination of *collapse* with the NP *sinkholes* (consisting of the prototypical noun *sinkholes*) in (5b) suggests that it is itself a NP (and a noun) and that the *to* is a preposition. Example (5c) concerns a virus strain. In such contexts, *drift* is a technical noun referring to a medical concept, i.e. the effect of cumulative mutations which eventually render a virus immune to vaccinations against its older incarnations, and this is what the article discusses. Lastly, (5d) involves the aspectual verb *keep* followed by a present participle. The doubl-*ing* constraint militates against the gerund in such environments, so it seems safest to treat such contexts as invariant and therefore ineligible for a quantitative analysis.

6.3   EXPLANATORY VARIABLES

[±Choice], Extraction, matrix modalization, matrix negation, matrix tense, and passivization were included in the analysis under the same rationale as in Chapter 5. Example (6a) involves [+Extraction] of a complement, while (6b) illustrates [+Choice], [+Negation] and [+Past], while (6c) exemplifies [+Modalization]. Example (6d) features [+Extraction] of an adjunct and [+Choice], while (6e) features passivization:

(6a)   You also can keep a "backup" box in your car that has duplicates of the things you are prone to losing ... (US, 2017, *Collective Evolution*)

(6b)   He was seen as a lone wolf on the council, not prone to compromise. (US, 2016, *Los Angeles Times*)

(6c)   Candidates whose recollections may be prone to ricochet should not apply, hence Mr Flynn's departure. (GB, 2017, *Herald Scotland*)

(6d)   Browning also looks into how some sectors of previously dominant male culture are prone to reacting to all these huge changes. (GB, 2016, *The Guardian*)

(6e)   Still, I have trouble envisioning him as a quiet kid prone to being bullied. (US, 2016, *Forward*)

Possible complementation differences between the two varieties were investigated as an obvious additional variable of interest. Based on the results obtained in the previous chapter, we also kept track of two variables that were found significant in the complementation with *accustomed*, i.e. the number of syllables in the stem of the complement verb, and postmodifier status of *prone*. This latter feature is seen in (6e) above. Due to the comparatively high semantic complexity of *prone* relative to the adjectives analyzed in the previous chapters, a number of additional variables of interest suggested themselves. They are discussed in the following.

6.3.1   *Animacy*

Subject animacy was treated as a quadrichotomy between Human, Other Animate, Concrete, and Abstract subjects. The assignment of these categories followed Zaenen et al. (2004). The Human category was reserved for humans and mythical humanoids (gods, ghosts, fairies etc.). The

Other Animate category includes animals (including bacteria), intelligent machines (such as computers and smartphones), vehicles, organizations, and personified inanimates, such as that seen in (7), a British example:

(7)   We all know by now that Game of Thrones isn't prone to doing anything by halves. (GB, 2015, *melty.com*)

Physical objects and substances perceptible to the five senses without the help of scientific equipment were treated as Concrete. This includes trees and body parts. The fourth category comprises events (e.g. the tournament), places (e.g. the Arctic), and abstractions (e.g. history).

6.3.2   Modal Scope

Another additional variable studied in this chapter is propositional scope. As a dynamic modal adjective with a secondary epistemic use, *prone* has much in common with the modal *may*, whose diachronic semantics are discussed by Bybee (1988). Like *prone*, *may* was originally restricted to a physical sense. The original meaning was approximately 'have the strength or power to'. This physical ability sense was first extended to the mental domain, then further generalized to refer to the presence of any enabling conditions, including those external to the subject. This generalization enabled the modal to acquire what has been called wide scope or propositional scope (Nordlinger and Traugott 1997: 301–304). With propositional scope, the modal no longer ascribes enabling conditions to a participant (exhibiting what Nordlinger and Traugott call "narrow scope"), but applies to the whole proposition. Once such a wide-scope dynamic meaning has been established, purely epistemic uses may arise through the conventionalization of the conversational implicature that if it enabling conditions exist for something to be the case (dynamic), then there is a non-zero probability that it is the case (epistemic). Many contexts allow either interpretation, and according to Bybee, such bridging contexts presumably play an important role when a dynamic modal is acquiring an epistemic sense. If the semantic development of *prone* is following a trajectory similar to *may*, then indeed its wide-scope dynamic and epistemic uses are innovative relative to the prototypical narrow-scope dynamic use, and such innovative use might conceivably influence complement choice.

Though unequivocally epistemic uses do occur (as in (1c) above), it is quite rare that a dynamic reading can be completely ruled out. More often, both readings are possible. In addition, since epistemic modals invariably have wide scope (Nordlinger and Traugott 1997: 304), it follows that classifying a given token of *prone* as epistemic is a stronger claim than merely ascribing it wide scope—purely epistemic tokens are a special case of wide-scope use.[3] We thus decided to use scope rather than epistemicity as an explanatory variable, assigning a score of 0 to tokens with unambiguously narrow scope, .5 to ones where either type of scope seemed possible, and 1 for unambiguously wide-scope uses which might or might not additionally be epistemic. Tokens exemplifying the three ratings are seen in (8a–c):

(8a)   The walls are so prone to falling they need to be propped up and stabilized by steel beams. (US, 2016, *The Boston Globe*) (narrow scope)

(8b)   Americans are becoming slightly less likely to be self-employed, and less prone to hold multiple jobs. (US, 2015, *Bloomberg View*) (narrow or wide scope)

(8c)   ... if you forget anyone while counting, you are prone to underestimating your true number of partners. (GB, 2015, *The Guardian*) (wide scope)

In (8a) it is clear that the risk of falling is attributed to the physical properties of the subject, so we are dealing with the dynamic use of *prone*. In (8b) the context is one of describing statistical trends, and it is not clear whether the intended meaning is one of a disposition toward multiple jobs that is becoming less common in the population (perhaps through demographic change), or whether *less prone* is intended to have propositional scope, with the approximate paraphrase 'holding multiple jobs is becoming less common among Americans'. It is especially worthy of note that the unambiguously epistemic adjective *likely* is used in the first coordinate, suggesting that *prone* may have been chosen in the second coordinate simply to convey the same epistemic meaning while avoiding repetition. In (8c) no inherent property or characterizing behavior is being attributed to the subject. Rather, *prone* has scope over the whole proposition, describing what is prone, or likely, to follow as a consequence if a particular error is committed.

6.3.3   Futurity

Futurity is the presence of a salient element of prediction in the context. Though wide-scope (and especially epistemic) uses of *prone* often concern speculation about what might happen in the future, propositional scope and epistemic modality are not limited to future reference. Conversely, sometimes *prone* retains its core semantic values of dynamic modality and characterization, i.e. narrow scope, even though the context of use clearly indicates that the characterization is being made due to its relevance for what might happen in the future. Example (9) illustrates:

(9)   Wednesday: Goldman Sachs, Citigroup, Netflix <p> Both Goldman and Citi are the most prone to going down after they report, Cramer said, because they tend to come in after reporting a good quarter. (US, 2017, *CNBC*)

This is a characterizing sentence that identifies recurrent patterns on the basis of past experiences. At the same time, the context is one of forecasting, and the characterization serves as the basis for an informed prediction on the following week's stock prices. A binary indicator was used to mark whether the context involved this kind of future orientation.

6.3.4   Stativity

A number of intuitions have been voiced in the literature regarding the semantic nuances of the variation between the infinitive and the gerund. Noonan (2007: 140) associates the gerund with factivity to such a degree that he stars *Zelda started sneezing but then didn't sneeze* as self-contradictory and indicates that grammaticality is restored by substituting a *to* infinitive. This is echoed by Krifka et al. (1995: 103), who point out that when referring to a specific past event, the verbal gerund in *Chewing tobacco calmed John down*, which is also a control construction, is hardly substitutable by a *to* infinitive. A very similar remark is also made by Huddleston and Pullum (2002: 1254). Allerton (1988) associates the infinitive with "a hypothetical eventuality" and "individual events", while associating the gerund with factuality, iterativity, completeness, and frequency. These notions are shared by Quirk et al. (1985: 1192–1193) and Langacker (2009: 300–301), who suggest that the -*ing* form may be preferred where the presence of a plural argument suggests multiple activities. Perhaps the boldest claim comes from Taylor and Dirven, who suggest that the *to* infinitive is obligatory after the aspectual verbs *begin* and *see* if the complement denotes a state (1991: 23–24).

In order to operationalize some of these intuitions, we first looked into coding each complement for the aspectual dichotomies of Dynamism/Stativity, Telicity/Atelicity, and Durativity/Punctuality. However, this proved difficult owing to the high degree of ambiguity involved. Verb phrases are often ambiguous with respect to at least one of the three features—lexical statives can be used dynamically in an inchoative sense; dynamic verbs can refer to telic or atelic events; and telic verbs can be punctual or durative (Smith 1997: 18, 58, 59). Arguments and adverbials can disambiguate, but often (especially with intransitives) there aren't any, and even if there are, they may themselves be ambiguous with respect to specificity and boundedness (see Lyons 1999 for discussion of the specificity and referentiality of NPs). This results in a high degree of subjectivity and arbitrariness in classification. To mitigate this problem, we discarded our seemingly unreliable classifications of Telicity and Durativity and limited ourselves to Stativity, which showed the most promise as a predictor as well as being the least difficult to classify. We applied the criteria of Lakoff and Ross (1966) and Smith (1997: 47, 184) to identify stative complements, again using a three-point scale with a score of 0 for dynamic complements (all previous data examples). Complements where either a stative or a dynamic interpretation was possible were rated .5, and unambiguous statives got a score of 1. These two types are illustrated by the examples below:

(10a)   He deemed people immutably wicked by nature and therefore prone to be hostile, a bias that brought his princes endless grief. (GB, 2015, *Telegraph.co.uk*)

(10b)   Study also shows that some breeds of dogs are more prone to being obese than other breeds. (US, 2016, *Parent Herald*)

Being hostile can be either stative (e.g. *The UK landscape is hostile to video shops.* GB, 2014, *The Guardian*) or dynamic (*These people are being hostile.* GB, 2010, *ChristianToday*). In main clauses, the progressive systematically triggers a dynamic interpretation (Smith 1997: 51–52), but the same contrastive distribution is not found with non-finite complementation of *prone*, which is characterized by free variation. Thus the complement of (10a), a *to* infinitive, may be either dynamic or stative—it is not clear whether the complement denotes a static quality or potential actions—while the complement of (10b) has a gerund despite the fact that the complement refers to a static quality (note the ungrammaticality of *Kim/the dog is being obese).

6.3.5   Repeatability

This variable was our attempt to operationalize Allerton's intuitions about the gerund's association with iterativity and frequency. Repeatability was rated at 1 if the complement situation was unmistakably of a kind that can recur virtually ad infinitum for a single entity. A score of .5 was assigned to complement situations that can technically recur for the same entity, but which are viewed as significant events with consequences, the relevant question being whether the situation occurs at all or not. Finally, a score of 0 was assigned to strictly non-repeatable complement situations. Examples (11a–c) from the American batch illustrate the three types:

(11a)   You still sleep with your phone by your side and you're prone to using your fave emoji or two. (US, 2017, *BuzzFeed News*) (repeatable)

(11b)   He blames the recent applications of the herbicide, which can be prone to vaporizing – or volatilizing – and drifting off-target. (US, 2017, *STLtoday.com*) (semi-repeatable)

(11c)   It finds a see-sawing pattern over that time in which sex is more prone to die in the womb, as various genetic influences take their toll. (US, 2015, *Phys.Org*) (non-repeatable)

Using emojis is infinitely repeatable and makes a good habitual predicate. The same is not true of vaporizing, for a dose of herbicide that has vaporized and drifted off target can surely be written off as a loss. However, it is technically possible for a substance to vaporize, condense back into a liquid, and then vaporize again with no limit to the number of such cycles. This contrasts with death, which is typically final and non-repeatable.

6.3.6   Grammatical Number

Note that the subjects in (11) are all singular or non-count. A hypothesis worth testing is whether a single, non-repeatable complement situation is likelier to be viewed as many situations (and favor the gerund) when the higher subject is plural—or more generally, whether plural higher subjects tend to favor the gerund relative to singular higher subjects. This was operationalized by coding each token for whether the higher subject was singular, non-count, a collective noun such as a *government*, or plural.

6.4   Statistical Analysis

6.4.1   Descriptive Statistics

Figure 6.1 shows univariate correlations between complement type and the variables described hitherto. As is immediately evident, all unmarked categories (and most marked ones) favor the gerundial variant. Indeed, the overall totals are 924 for the gerund and 576 for the infinitive. As can be expected by now, Extraction contexts represent a conspicuous exception to the overall predominance of the gerund. The same seems to

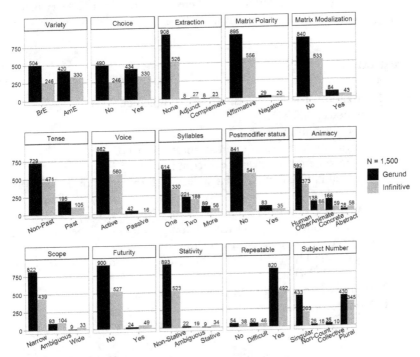

Fig. 6.1   Marginal distributions of the explanatory variables by complement type

hold true for abstract matrix subjects, futurity, stative complements, and propositional scope. Worth noting is also the ostensibly more pronounced gerund-favoring behavior of British English relative to its upstart colonial counterpart. In the next section, we inspect whether these associations persist when every variable is considered simultaneously.

6.4.2   Multivariate Analysis—Model Selection

As usual, we utilized the *lme4* package in R to fit mixed-effects logistic regression models with the lemma of the complement verb as the sole random effect. This statistical method is outlined in Sect. 2.3.3. We again followed the model-selection procedure detailed in Sect. 2.3.4—starting from a maximal main-effects model which included all the variables described in Sect. 6.3, we manually pruned the model of non-significant predictors in a stepwise fashion, checking for confounding at each step as well as testing each quantitative predictor for non-linear effects before keeping or removing it definitively. The initial model is summarized below.

Fixed Effects:

1. Extraction (trichotomous: None, Complement, Adjunct)
2. [±Choice] (dichotomous)
3. Voice (dichotomous)
4. Stativity of complement (three-point ordinal: 0, 0.5, 1)
5. Future-orientation of *prone* (dichotomous)
6. Repeatability of complement situation (three-point ordinal: 0, 0.5, 1)
7. Matrix negation (dichotomous: None, *Not*-negation)[4]
8. Subject animacy: (quadrichotomous: Human, Other Animate, Concrete Inanimate, Non-Concrete Inanimate)
9. Grammatical number of subject (quadrichotomous: Singular, Non-Count, Collective, Plural)
10. Syllable count of subordinate verb (quantitative)
11. Modal scope (three-point ordinal: 0, 0.5, 1)
12. Matrix modalization (dichotomous: unmodalized higher clauses vs ones involving a modal verb or adverb)
13. Matrix tense (dichotomous: Non-Past i.e. present or perfect vs Past i.e. preterite or preterite perfect)
14. Postmodifier status of *prone* (dichotomous)

15. Variety (dichotomous: BrE vs AmE)

Random Effects:

1. Subordinate verb (nominal-scale with 581 categories)

The variance inflation diagnostics detected no significant multicollinearity among the explanatory variables. Voice, matrix modalization, matrix tense, and postmodifier status of the head were all diagnosed as statistically and predictively non-significant and dropped from the model in the ensuing backward-elimination process. Perhaps surprisingly, the same conclusion had to be drawn regarding matrix negation, which had a negligible coefficient of 0.21 ($G = .42$; $p = .52$). Given the results of Chapters 2, 4 and 5, we checked for an interaction between this variable and [±Choice] before proceeding. Sure enough, the results pointed to a pattern directionally similar to what was observed in all the previous datasets. Negated matrices were estimated to favor *to* infinitives by an odds ratio of $e^{0.78} = 2.2$ in [−Choice] contexts but not in [+Choice] contexts ($e^{0.78−0.90} = 0.88$). However, the interaction term was statistically non-significant, ($G = 1.8$; $p = .18$). *Prone* appears to be rarely negated in general—our total 1,500 observations included only 49 negated tokens. This scarcity of data prevents confident inferences about the effect that the negation of *prone* may exert on its complement selection. We therefore removed matrix negation from the model.

In addition to the aforementioned exclusions, some of the remaining explanatory variables had to be recoded in the interest of model fit and parsimony. Firstly, complement and adjunct extraction had near-identical coefficients around 1.6, so they were subsumed under a single category. Second, the important distinction in subordinate verb length turned out again to be that between mono- and polysyllabic verbs, so we dispensed with linear syllable counts in favor of a binary indicator of whether the lower verb stem had more than one syllable. Third, it proved advantageous to dichotomize the grammatical number classification of matrix subjects, contrasting only a marked category of non-count and plural subjects against a reference category comprising singular and collective subjects. Fourth, the difference between human and collective or anthropomorphized subjects turned out to be statistically superfluous, so we replaced the quadrichotomy with a trichotomy of human/animate, concrete inanimate, and non-concrete inanimate. Finally, goodness-of-fit considerations necessitated that the three-point numerical scale employed to quantify Repeatability be abandoned in favor of a dichotomous scheme contrasting easily repeatable complement situations with ones whose repetition is difficult, unlikely, or impossible.

After completing these model simplifications, we checked algorithmically for statistically significant interactions. Three were identified. The strongest interaction was found between futurity and polysyllabicity of the complement verb, suggesting that futurity exerted a large infinitive-favoring effect with an odds ratio of $e^{1.64} = 6.15$, but only for monosyllabic verbs. The second interaction suggested that polysyllabicity of the complement verb was only significant in [+Choice] contexts. We deemed both of these interactions implausible, proceeding to dismiss the associated models. The third interaction was detected between Variety and EasyRepeat, suggesting that easily repeatable complement situations favored the gerund more strongly in AmE than in BrE. This is not implausible, however, given that the difference was quantitative rather than qualitative, the simpler model seemed preferable. In particular, even though adding the interaction term improved goodness-of-fit as measured by the likelihood-ratio test ($G = 3.89$; $p = 0.049$), it weakened predictive power as measured by the concordance index, which decreased from .766 to .764. Therefore, we opted for the more parsimonious model without interactions, given that it was simpler to interpret and performed essentially as well (see Agresti 2018: 130 for a similar example). The final model contains the following explanatory variables.

Fixed Effects:

1. Extraction (dichotomous)
2. [±Choice] (dichotomous)
3. Stativity of complement (three-point ordinal: 0, 0.5, 1)
4. Future-orientation of *prone* (dichotomous)
5. Repeatability of complement situation (three-point ordinal: 0, 0.5, 1)
6. Subject animacy: (trichotomous: Animate, Concrete Inanimate, Non-Concrete Inanimate)

7. Grammatical number of subject (dichotomous: Singular/Collective, Non-Count/Plural)
8. Polysyllabicity of subordinate verb (dichotomous)
9. Modal scope (three-point ordinal: 0, 0.5, 1)
10. Easy repeatability of complement situation (dichotomous)
11. Variety (dichotomous: BrE vs AmE)

Random Effects:

1. Subordinate verb (nominal-scale with 581 categories)

The next section interprets the model output.

6.4.3 Multivariate Analysis—Model Interpretation

Figure 6.2 has the estimated coefficients and their 95% profile-likelihood confidence intervals, which imply statistical significance at the .05 level when located entirely on one side of zero. The model achieves a simple classification accuracy of .715 and a concordance index of .766. These scores are lower than those reported for the previous models discussed in this book. Notably, concordance index values between .7 and .8 are considered only "acceptable" discrimination according to Hosmer et al. (2013: 177). The first thing to suspect is that there may be crucial predictors of this alternation that remain undiscovered. Though that is by no means implausible, it should be noted that the present analysis has investigated a larger set of explanatory variables than any of the previous chapters, and the number explanatory variables included in the final model is also greater than in previous chapters. It is therefore far from clear whether we can conclude that important variables are missing. A second possible explanation, not mutually exclusive with the first, is that the strongest predictors have imbalanced distributions in the dataset. Indeed, a brief glance at Figs. 6.1 and 6.2 reveals that the most strongly predictive features—stative complements, extractions, unambiguously wide scope, and non-concrete inanimate subjects—are very infrequent in the dataset. In other words, their predictive value applies in very few cases. This goes a long way to account for the comparatively low predictive performance of the model.

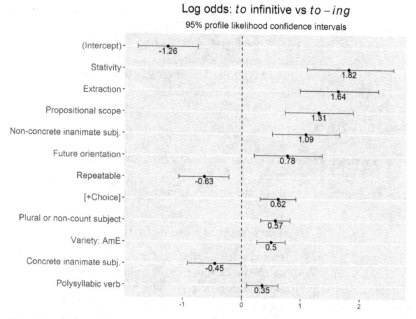

Fig. 6.2 Point estimates for all statistically significant predictors on the logit scale

The Intercept reflects that in a context with every variable at its "default" value, i.e. a complement featuring no extraction or future reference and denoting a hard-to-repeat non-stative situation encoded by a monosyllabic verb (e.g. *die*) with an unagentive singular animate subject in British English, the odds of the *to* infinitive are estimated at $e^{-1.26} = 0.28$ (corresponding to a probability of 22%). For an analogous observation in American English, these odds multiply by a ratio of $e^{0.5} = 1.65$. Consistent with previous findings, extraction contexts favor the infinitive at an odds ratio of $e^{1.64} = 5.13$ compared to analogous contexts with no extraction. Also consistent with previous findings, [+Choice] contexts increase the odds of the *to* infinitive by a factor of $e^{0.62} = 1.86$ relative to analogous [−Choice] contexts. The apparent predilection of polysyllabic verbs for the infinitive variant, first pointed out by Rohdenburg for *accustomed* (2016: 474–475), is likewise congruous with previous results. By contrast, postmodifier status of *prone* proved non-significant in the present dataset, its estimated coefficient barely 50% of the one observed in the previous chapter, albeit with the same sign.

Regarding the additional variables, Stativity emerges as a key predictor, with an estimated odds ratio of $e^{1.82} = 6.15$ in favor of the infinitive. Propositional scope and abstract higher subjects show a considerable preference for the infinitival variant as well, with estimated odds ratios of $e^{1.3} = 3.69$ and $e^{1.09} = 2.97$, respectively. By contrast, concrete inanimate subjects differ dramatically from abstract ones, with an apparent tendency to attract the gerund at an odds ratio of $e^{0.45} = 1.57$ compared to the default category of animate subjects, and at $e^{0.45+1.09} = 4.68$ compared to abstract subjects.[5] Insofar as concrete subjects represent an intermediate category between animate and abstract ones, such a result may be viewed as unexpected. More unexpected, however, is the effect of the grammatical number of the subject. Contrary to our hypothesis, mass and plural subjects show a moderate but highly statistically significant preference for the infinitive ($e^{0.57} = 1.77$; $W = 4.5$; $p < .001$), and this tendency is shared by mass noun subjects.

In accordance with our predictions, repeatability of the complement situation turns out to be a factor favoring the gerundial variant. The salient distinction appears to be between "easy" and "difficult" repeatability, rather than whether recurrence is logically possible or impossible. Complement situations of "difficult" and zero repeatability favor infinitives at an odds ratio of $e^{0.63} = 1.88$.

Finally, the lemma of the complement exerts only a minor influence on the variation. The standard deviation of the random effect is 0.43, indicating that the average verb is estimated to favor one or the other variant at an odds ratio of $e^{0.43} = 1.53$ relative to the population mean. Expressed as an intracluster correlation, this implies that lemma-specific idiosyncrasies account for only $0.43^2/(\pi^2/3 + 0.43^2) = 5.3\%$ of the variation. The random effect is statistically significant at the .05 level ($G = 3.29$; $p = .07/2 = .035$), but it should be noted that this is the smallest random effect in any of the five analyses conducted in this book—its statistical significance is largely due to the size of the dataset.

6.5   DISCUSSION

Most of these results are easily explained by viewing them against the "prototypical" use of *prone* as a characterizing dynamic modal. The prototypical use involves a subject that is tangible, concrete, and has the requisite degree of stability across time. Concreteness facilitates the attribution of inherent properties, while temporal stability facilitates the attribution of regular patterns of behavior. Regularity can only be established by repeated attestations of the pattern or behavior, so another prototypical feature is a complement situation that is repeatable. Strict technical or logical repeatability is not enough, however—a prototypical complement situation is a repeatable one that makes a good characterizing predicate, as in *students (are prone to) get(ting) bored at long lectures*. Since the prototypical use of *prone* involves repeated past occurrences, the gerund—with its associations of iterativity and specificity—is a natural match for it.

If this is correct, it follows that the absence of any of the aforementioned prototypical semantic features decreases the selection odds of the unmarked, prototypical complement, i.e. the prepositional gerund. Abstract subjects repel the prepositional gerund because they make bad possessors of inherent properties. Non-repeatable situations repel the prepositional gerund because they make bad characterizing predicates. Future prediction repels the prepositional gerund because it seldom involves characterization. Propositional scope repels the prepositional gerund because the characterization does not involve a subject at all, but a proposition. In fact, instances of *prone* with propositional scope are analogous to those with abstract subjects in the sense that both phenomena involve the attribution of a tendency or inclination to something that lacks concrete, cognitively simple, and straightforwardly identifiable real-world referents. It is therefore not clear whether the respective effects of abstract subjects and propositional scope should be attributed to the incompatibility of these features with the prototypical semantics of *prone* or to their relatively high level of cognitive abstractness, i.e. the Complexity Principle. Lastly, stative complements repel the gerund because states are conceptualized as static, so they are incompatible with the frequency and iterativity associations of the gerund. This finding is not unlike that of Cuyckens et al. (2014: 196), who find stative complements to prefer finite clauses to non-finites in the complementation of *remember, regret,* and *deny*. The analogy is that just as finite clauses are more explicit than non-finite ones, *to* infinitives are more explicit than prepositional gerunds (Rohdenburg 2016: 472).

While the aforementioned cluster of predictive semantic factors may well be due to the core lexical semantics of *prone*, the same is not true of the three factors [±Choice], Extraction, and polysyllabicity of the lower

verb, which seem to be of quite general applicability in English grammatical variation. In accordance with previous work on the Complexity
Principle, the latter two features have now been shown to favor the
infinitive in the clausal complementation of *prone*.

The Choice Principle has been found significant with numerous other
adjectives (including the ones discussed in earlier chapters of this volume),
and an unmistakable probabilistic proclivity for the *to* infinitive is also
apparent in agentive complements of *prone*. The Choice Principle also
provides a plausible explanation for the finding that concrete inanimate
subjects seem more favorable to *to* -*ing* than animates. Before subsuming
passive complements under the superordinate category of [−Choice]
complements in the model-selection process, we noticed that their statistically non-significant coefficient was favorable to the gerund. This is
in accordance with the view that passive complements are particularly
clearly [−Choice]. We suggest that the same is true of concrete inanimate
subjects. This contrasts with animate subjects, where many [−Choice]
complements retain a tinge of agentivity:

(12a) While I may have misspoken on rare occasions – as anyone
       who is human is prone to err – my loyalty to Dr. Carson and
       his candidacy is clear and unwavering... (US, 2016, *Wall Street
       Journal*)
(12b) Now a physicist believes he has unravelled why mobile phones
       seem so prone to landing screen side down when they are
       dropped. (GB, 2015, *Daily Mail*)

Many of the mishaps befalling human subjects, though often unintended
and unagentive, would have been avoidable through greater caution.
The same is not true for concrete inanimate subjects. Unless they are
anthropomorphized (and hence no longer strictly inanimate), they are
paradigm examples of patient subjects. This may suffice to explain their
gerund-favoring behavior relative to animates.

It is more difficult to explain why plural and non-count subjects seem
to favor *to* infinitives. In particular, we expected plural subjects to favor *to*
-*ing* due to the hypothetical association of a multitude of referents with
a multitude of events. This expectation has proven false. In theory, one
might argue that plural and mass subjects are marked relative to singular
ones, so that the result might be simply a manifestation of the Complexity

Principle at a very basic level. Whether this suggestion is on the right track is naturally subject to further investigation.

Lastly, the multivariate analysis has confirmed what the descriptive statistics already suggested in Sect. 6.4.1, i.e. that American English is more favorable to the infinitival variant than British English.

6.6   CONCLUSION

This chapter has investigated non-finite alternation under the adjective *prone* in British and American English. We have included a host of old and new syntactic and semantic variables in the multivariate analysis and shown that important predictors of the complementation of *prone* exist in both the syntactic and the semantic domain. The Extraction Principle and polysyllabicity of the complement verb have once more proven to be significant predictors, lending further support to Rohdenburg's Complexity Principle. It has also been observed again that it is of minor importance in the choice of non-finite complement whether the gap left by extraction from the lower clause corresponds to a complement or to an adjunct. Among semantic factors, the Choice Principle has likewise emerged as significant once more.

As for the cluster of additional semantic variables identified as significant, we have interpreted them against the background of a primary, prototypical use of *prone* as a characterizing dynamic modal adjective. Such an adjective, we propose, combines most naturally with subjects referring to concrete entities with temporally stable inherent properties to which speakers can attribute behaviors that are regular or recurrent, and thus a natural match for the prepositional gerund. Broadly speaking, our interpretation is that the more the context and communicative intent differs from these prototypical features, the likelier the selection of the infinitival variant.

Though many statistically significant predictors have been identified—some of them with impressive coefficients—the overall explanatory power of the model remains lower than what was achieved by our models for the complementation of *afraid* and *accustomed*. This is despite the fact that our model for *prone* includes the largest number of significant predictors of any statistical model presented in this volume. Part of the reason is indubitably the low in-sample frequency of the strongest predictive features, resulting in a proportionally small collective contribution to the model's overall explanatory power in-sample. Nonetheless, it may well be that there remain key predictors that we have failed to uncover. One fruitful place to look for missing predictors might be Smith's aspectual features of Telicity and Durativity, a course which we briefly pursued but had to ultimately abandon due to the formidable challenges posed by the multiple sources of aspectual ambiguity that frequently combine to render conclusive classification of corpus examples an arduous and error-prone endeavor.

We should bear in mind, however, that *prone* exhibits a much higher degree of micro-polysemy and semantic versatility than *afraid* and *accustomed*, which are limited to animate subjects. Although it seems clear that the prototypical sense of complement-taking *prone* is characterizing and dynamic, there is probably variation in just how salient its secondary, wide-scope/epistemic sense is in different speakers' mental lexicons. Conceivably, the wide-scope/epistemic sense may pattern alike with other epistemics such as *bound*, taking mostly or only infinitival complements, and this use may be the dominant one for a certain minority of speakers. If this is the case, then much of the unexplained non-finite alternation under *prone* could be simply due to inter-speaker variation in whether the prototypical, dynamic use or the wide-scope/epistemic use is felt to be the primary one. Then there would be little hope of accurately modeling the variation unless speaker information was available on every token.

However, regardless of whether additional factors bearing on the complementation of *prone* are ultimately identified, ascertaining whether the explanatory variables discovered by the present study are useful predictors of syntactic variation with other complement-taking predicates should be a rewarding and interesting enterprise.

NOTES

1. *Bent* is another adjective with a primary sense referring to a physical posture which also takes sentential complements (introduced by *on*) denoting actions that the subject is disposed to take.
2. We are again indebted to Mark Davies, who granted permission for the first author to share concordances from the local copy of this corpus with the second author.
3. The difference between purely epistemic and wide-scope dynamic uses is that in the former case, the sense of characterization and temporally stable enabling conditions is absent—rather, the focus is strictly on whether something is likely to be or come true or not, as in (1c) above.

4. *No*-negation of the higher clause was present in only three data points of the preliminary sample. Since this was far too few to enable any meaningful statistical inference, we set the tokens aside.
5. Changing the sign of the coefficient yields an analogous interpretation of the effect in which the respective roles of the success and failure outcome are reversed.

References

Allerton, David. 1988. Infinitivitis in English. In *Studies in Descriptive Linguistics (Essays on the English Language and Applied Linguistics on the Occasion of Gerhard Nickel's 60th Birthday)*, ed. Josef Klegraf and Dietrich Nehls, 11–23. Heidelberg: JuliusGroos.

Agresti, Alan. 2018. *An Introduction to Categorical Data Analysis*, 3rd ed. Hoboken, NJ: Wiley.

Bybee, Joan. 1988. Semantic Substance vs. Contrast in the Development of Grammatical Meaning. Proceedings of the 14th Annual Meeting of the Berkeley Linguistics Society: 247–264.

Cuyckens, Hubert, Frauke D'hoedt, and Benedikt Szmrecsanyi. 2014. Variability in Verb Complementation in Late Modern English: Finite vs. Non-finite Patterns. In *Late Modern English Syntax*. Studies in English Language, ed. Marianne Hundt, 182–204. Cambridge: Cambridge University Press.

Hosmer, David W., S. Lemeshow, and Rodney X. Sturdivant. 2013. *Applied Logistic Regression*, 3rd ed. Hoboken, NJ: Wiley.

Huddleston, Rodney D., and Geoffrey K. Pullum. 2002. *The Cambridge Grammar of the English Language*. 3rd printing 2010. Cambridge: Cambridge University Press.

Krifka, Manfred, Francis J. Pelletier, Greg N. Carlson, Alice Ter Meulen, Gennaro Chierchia, and Godehard Link. 1995. Genericity: An Introduction. In *The Generic Book*, ed. Greg N. Carlson and Francis Jeffry Pelletier, 1–124. Chicago, IL: University of Chicago Press.

Lakoff, George, and John R. Ross. 1966. A Criterion for Verb Phrase Constituency. In *Mathematical Linguistics and Automatic Translation*, 1–11. Cambridge, MA: Harvard University.

Langacker, Ronald W. 2009. *Investigations in Cognitive Grammar*. Berlin and New York: Mouton de Gruyter.

Lyons, Christopher. 1999. *Definiteness*. Cambridge and New York: Cambridge University Press.

OED Online. 2019. Oxford University Press, June. www.oed.com/view/Entry/152519. Accessed 21 August 2019.

Noonan, Michael. 2007. Complementation. In *Language Typology and Syntactic Description, Second Edition: Volume II: Complex Constructions*, ed. Timothy Shopen, 52–150. Cambridge and New York: Cambridge University Press.

Nordlinger, Rache, and Elizabeth C. Traugott. 1997. Scope and the Development of Epistemic Modality: Evidence from Ought To. *English Language and Linguistics* 1 (2): 295–317.

Palmer, Frank R. 2001. *Mood and Modality*, 2nd ed. Cambridge: Cambridge University Press.

Poutsma, Hendrik. MS. *Dictionary of Constructions of Verbs, Adjectives, and Nouns*. Unpublished. Copyright Oxford University Press.

Quirk, Randolph, Sidney Greenbaum, Geoffrey Leech, and Jan Svartvik. 1985. *A Comprehensive Grammar of the English Language*. London: Longman.

Rohdenburg, Günter. 2016. Testing Two Processing Principles with Respect to the Extraction of Elements out of Complement Clauses in English. *English Language and Linguistics* 20: 463–486.

Ross, John R. 1972. Doubl-ing. *Linguistic Inquiry* 3 (1): 61–86.

Smith, Carlota S. 1997. *The Parameter of Aspect*, 2nd ed. Dordrecht and Boston: Kluwer.

Taylor, John, and René Dirven. 1991. *Complementation*. Duisburg: L.A.U.D. Linguistic Agency, University of Duisburg.

Zaenen, Annie, Jean Carletta, Gregory Garretson, Joan Bresnan, Andrew Koontz-Garboden, Tatiana Nikitina, Catherine M. O'Connor, and Tom Wasow. 2004. Animacy Encoding in English: Why and How. Proceedings of the 42nd Annual Meeting of the Association for Computational Linguistics (ACL'04), Workshop on Discourse Annotation, pp. 118–125. Barcelona, Spain.

CHAPTER 7

Conclusion

Abstract This chapter presents an overview of some of the main findings of the book. The book focuses on explanatory factors bearing on the alternation of *to* infinitival complements with gerundial patterns—the *of* -*ing* pattern in the case of the adjective *afraid* and the *to* -*ing* pattern in the case of the adjectives *accustomed* and *prone*. When large corpora have become available in recent years, such factors have been the object of intense investigation on the basis of univariate analysis, but this book makes essential use of multivariate analysis, and is thus able to provide a clearer understanding of the impact of each individual factor on complement selection, as well as an insight into the interaction of the different factors. It is also seen that the significance of one and the same explanatory factor can vary, depending on the construction in question. For instance, the Choice Principle, proposed in very recent work, is found to have a dramatic effect in the case of *afraid*, but only a moderate one in the case of *accustomed*. The chapter also outlines some research questions arising from the findings of the book for further work in the study of non-finite complementation.

Keywords Infinitival complements · Gerundial complements · Multivariate analysis and its application

© The Author(s) 2021
J. Ruohonen and J. Rudanko, *Infinitival vs Gerundial Complementation with Afraid, Accustomed, and Prone,*
https://doi.org/10.1007/978-3-030-56758-3_7

149

The area of sentential complementation is at the interface of syntax and semantics, and it offers a rich array of tasks for investigating grammatical variation in recent English. This book offers a number of case studies on variation in cases where a single head selects more than one type of sentential complement, and the focus is on selected matrix adjectives that select both *to* infinitival and gerundial complements such that the resulting constructions are fairly close to each other in meaning. One of the gerundial constructions is the *of -ing* pattern, which was studied with respect to the complementation of the adjective *afraid*, and the other gerundial pattern is the *to -ing* pattern, which was studied with respect to the complementation of the adjectives *accustomed* and *prone*. The methodological approach of each content chapter is largely corpus-based. While our results are based directly on the corpora used, we believe them to have a degree of generalizability to the national varieties that the corpora represent. More specifically, we believe that the language-internal (i.e. semantic and syntactic) variables identified as significant operate below the level of conscious awareness, and we deem it unlikely that their effects would be directionally different in new datasets representing different configurations of language-external variables within the same variety.

The most important contribution of the present book to the research field is probably its systematic use of multivariate regression analysis, which quantifies the independent effects of the explanatory variables on the outcome, holding constant all the other explanatory variables included in the statistical model. The main focus of the multivariate analyses is on explanatory variables pertaining to the domains of syntax and semantics. Most of the syntactic explanatory variables investigated in this book are relevant from the standpoint of the so-called Complexity Principle (Rohdenburg 1996), which predicts that the presence of features which increase processing complexity will favor the selection of infinitival complements over gerundial complements. Within the domain of semantics, the main focus of the present work has been on the Choice Principle, which proposes that the *to* infinitive tends to be favored over gerundial complements when the understood subject of the lower clause has the thematic role of Agent.

With respect to the complementation of the adjective *afraid*, the study that the present authors completed first concerns non-finite variation in the Strathy Corpus and the BNC, and Chapter 2 opens with this investigation. The explanatory variables of the model here included some that have already achieved an established status in the literature, including the Extraction Principle, whose validity could not be reliably assessed without larger corpora. They also included the Choice Principle. The principle has been set up on the basis of univariate analysis in very recent work and it is recalled that in the Introduction the research question was tabled whether the Choice Principle can stand the test of multivariate regression analysis. The data from Strathy and the BNC yielded a positive answer to this question, providing corroboration for the principle. It is worth adding that in both datasets the Choice Principle was a significant factor impacting complement selection even after voice was singled out as a separate explanatory variable, which had not been the case in the univariate analysis studies where the principle had first been proposed. Passive complements constitute an environment in which the Complexity Principle and the Choice Principle are in conflict. Passivization is known to increase processing load, but the subjects of passive clauses also tend to be prototypical instantiations of the Patient role. The fact that no infinitive-favoring (nor disfavoring) effect was detected for passive complements after controlling for [±Choice] suggests that semantics may outrank syntax with respect to complement choice.

Regarding text type, a statistically significant infinitive-favoring effect was found in the Canadian data for the Fiction genre, while the British data showed a gerund-favoring effect for the more bookish registers of Academic and Learned Prose. While these findings must be viewed as provisional, they are hardly surprising given that one would expect fictional narratives to be mostly concerned with specific complement situations, whereas learned prose presumably concerns itself with generalizations; generalizations have been proposed in previous literature as one of the connotations of the *of -ing* complement (Rudanko 2015: 40). As regards other variables, extraction contexts proved to be too few to properly investigate in corpora this small—at least in the complementation of *afraid*—so their analysis had to be deferred to subsequent chapters.

Chapter 3 took up a phenomenon that had not been discussed in previous literature, namely the multitude of different constellations of infinitival clauses that can occur when an adjective of fear is premodified by *too*. We noted the fascinating syntactic indeterminacy that may arise in such contexts, proposing one semantic and two syntactic criteria to help resolve it. The semantic criteria involve the time relations between the situations described by the two infinitival clauses, while the syntactic criteria are based on meaning-preserving syntactic transformations. We feel that the proposed criteria are helpful in ascertaining which infinitival clauses are adjuncts and which ones complements, as well as in determining which clause is complementing which head. The chapter closed with speculation on a potential gerund-favoring effect in the complementation of fear adjectives when the adjective is nested within a degree-complement construction introduced by *too*, which characteristically selects an infinitival complement that is to follow the eventual non-finite complement of the adjective.

Chapter 4 tied up the loose ends left by the two previous ones, employing a purposeful sampling technique in conjunction with a massive corpus in order to inquire into the statistical effects of several rare grammatical features on the non-finite complementation of *afraid*. The rare features of interest included nesting of the adjective within a degree-complement construction, extraction contexts, potential *horror aequi* contexts, complement negation, and insertions. The Choice Principle and matrix negation were also investigated, and the former was once again confirmed in that lower subjects of the agentive type were shown to exert a dramatic infinitive-favoring effect. As in Chapter 2, passivization of the complement was found immaterial to variant selection when [±Choice] has already been taken into account.

As for the rare syntactic features, the multivariate analysis revealed the error in our intuitions about the effect of nesting within a degree-complement construction, as the model showed that such nesting had no additional gerund-favoring effect after adjusting for [±Choice]. By contrast, a strong and statistically significant effect was found for insertions, as predicted by Vosberg (2003: 210–211). An interesting, if not surprising, new finding was that insertions between the adjective and its complement had almost twice the infinitive-favoring odds ratio of insertions within the complement. As for *horror aequi*, a striking dispreference was found for two *of* -*ing* clauses separated only by *afraid*, while no statistically significant avoidance of two near-contiguous *to* infinitives could be confirmed. This was attributed to the greater phonological bulk of the prepositional gerund relative to the infinitival variant. Finally, the larger sample enabled us to obtain robust backing for the findings that had previously been observed for extraction contexts in univariate analyses, i.e. a strong preference for the *to* infinitive.

In Chapter 5, we shifted focus to a slightly different non-finite alternation, namely that between *to* infinitives and the gerundial *to* -*ing* variant under the adjective *accustomed*. One appealing feature of this adjective relative to *afraid* was its fairly high rate of co-occurrence with extraction contexts. This enabled us to analyze whether the results obtained for extraction contexts through purposeful sampling of the adjective *afraid* accorded with the results seen in the more naturalistic sample of extractions involving *accustomed*, which fortuitously accompanied the dataset collected from the mid-twentieth century, i.e. the period we had identified as the time of most significant variation between the two complement types.

The result seen for Extraction in Chapter 4 was indeed corroborated in the *accustomed* dataset in Chapter 5. Similarly consonant with the previous chapter were the findings on pre-complement insertions, which strongly favored *to* infinitives with *accustomed* as well. The results on other variables relating to form were less conclusive. There was no statistically significant *horror aequi* effect on the incidence of *to* infinitives as complements of heads preceded by a *to* infinitive; although the coefficient was large and had the expected sign, there were too few relevant tokens in the 1,400-strong dataset to support reliable inference. *Aequi* contexts with an immediately preceding *to -ing* could not be analyzed at all due to their scarceness. This shows that such rare features are hard to investigate without the kind of purposeful sampling conducted in Chapter 4, except in datasets of great size. Two form-related variables showing some promise of predictive utility were the syntactic postmodifier status of the adjective and the syllable count of the lower verb. The former feature was found to favor the gerundial variant, while phonologically heavier lower verbs were seen to subtly favor *to* infinitives.

Regarding semantic variables, the Choice Principle proved to be a statistically significant infinitive-favoring factor once again, although its effect was much less dramatic than what was observed with *afraid* in earlier chapters.

Chapter 6 turned to the most semantically complex adjective treated in this volume. *Prone*, which we described as a characterizing modal adjective, differs from *afraid* and *accustomed* in that it is not restricted to animate predicands, but is also used in characterizing inanimates, abstractions, and propositions. The non-finite complementation of *prone* today, like that of *accustomed* in the mid-20th century, is characterized by a more or less free variation between the *to* infinitive and gerundial *to -ing*. The multifaceted nature of this adjective is reflected in the fact that although the analysis considered a larger number of explanatory variables than any of the preceding ones, the classification and discrimination accuracy of the resulting statistical model was appreciably lower than what we had accomplished with the previous adjectives.

Featuring the largest dataset analyzed in this volume, our sample of 1,500 observations was evenly split between British and American English in order to facilitate a comparison between the two highest-profile national varieties of the language in addition to the usual analysis of the role of syntactic and semantic variables in complement choice. American English turned out to be more hospitable to the infinitival variant overall than its matrilect. This matches what was observed for *afraid* in Chapter 2, where Canadian English—another New World variety—was found to exhibit a significantly heavier predilection for the infinitival variant than its mother dialect.

With respect to semantic factors influencing the alternation, the Choice Principle was once again statistically significant, but the effect was once again much more modest than that seen in the complementation of *afraid*. As far as other semantic factors were concerned, the most dramatic effects were found for stativity of the complement situation and the animacy category of the lower subject. Abstract lower subjects were found to strongly favor infinitival complementation, while concrete inanimates were found to subtly favor *to -ing* even after controlling for agentivity, perhaps because concrete inanimates are particularly paradigmatic examples of the Patient role. Possibly related to abstractness of the lower subject was the finding that when *prone* had propositional scope, modalizing the entire proposition rather than attributing a tendency to a subject, infinitives were favored in much the same way as with abstract lower subjects. Contexts involving a salient element of future prediction were identified as another infinitive-favoring circumstance, while complement situations classified as easily repeatable were found to favor the gerundial variant.

Apart from the Choice Principle, which may be a general tendency in the alternation between infinitival and gerundial complements, we suggested that most of the semantic features found to explain non-finite alternation under *prone* have an explanation when contrasted with its prototypical use as a characterizing adjective that ascribes recurrent patterns of behavior to concrete entities. This hypothesized prototypical use was associated with *to -ing*, since the literature seemed unanimous about the connotations of recurrence and durativity inherent in the gerund. All semantic features deviating from the prototypical characterizing use would then be expected to increase the probability of the infinitival variant. The only semantic factor that we struggle to explain is the observed tendency for plural and non-count complement subjects to favor the infinitive. Attributing the phenomenon to Rohdenburg's Complexity Principle on account of the hypothetically higher semantic complexity of plural and non-count nouns relative to singular count nouns may be a possible explanation, but the question will deserve further investigation.

As regards factors related to form, polysyllabic verbs were again found to exhibit a slight but statistically significant preference for infinitival complements. More importantly, however, extraction contexts were again corroborated as an important factor favoring *to* infinitives, and there seemed to be no difference between complement extraction and adjunct extraction from the standpoint of non-finite alternation.

The most consistently significant predictors of non-finite complement choice throughout the chapters have been the Extraction Principle and the Choice Principle. The former principle already has a well-established status in the literature. The Choice Principle, by contrast, is very new and less well established, and a noteworthy result of the present work has been to show that the agentivity of the lower subject is a significant predictor even when adjusting for passivization. Passivization, on the other hand, has consistently tested non-significant when adjusting for [±Choice]. However, one should also bear in mind the considerable difference in the effect size of the Choice Principle between *afraid*, which alternates between *to* infinitives and *of* -*ing*, and the other two adjectives, for which the gerundial variant involves the preposition *to*. The [+Choice] effect is dramatic for *afraid*, with an odds ratio in the double digits in all three datasets, while for *prone* and *accustomed* it is better characterized as a mild probabilistic bias. The crucial question is then whether the difference is due to the semantics of the adjective or the identity of the preposition. More specifically, does the relationship between complement choice and contextual semantic factors such as [±Choice] differ for what may broadly be characterized as "disinclination" adjectives such as *afraid*, and "inclination" adjectives such as *accustomed* and *prone*? This seems a fairly plausible hypothesis. The alternative, lexis-based hypothesis is that the difference is due to the identity of the preposition (*of* vs *to*) of the gerundial variant. To determine which hypothesis is correct, one line of future investigation would be to undertake a multivariate analysis comparing the non-finite complementation, on the one hand, of two

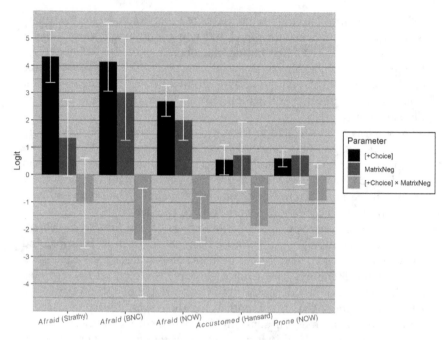

Fig. 7.1   Parameter estimates and 95% profile-likelihood confidence intervals for [+Choice], matrix negation, and their interaction in the five datasets analyzed in this book

or more near-synonymous heads for which the preposition of the gerundial complement was different, and of two or more semantically dissimilar heads for which the preposition of the gerundial complement was the same.

Matrix negation has been identified as statistically significant in three of the five analyses. Statistically significant or not, however, its role in complement selection appears to be directionally similar in every dataset. That role seems to involve an interaction with the Choice Principle. Figure 7.1 plots the estimated effects of both variables and their interaction, along with 95% profile likelihood confidence intervals, in the five analyses conducted. The general pattern is the same in all—with other things equal, negated matrices attract *to* infinitives in [−Choice] contexts but not in [+Choice] contexts. The infinitive-favoring effect of negated higher clauses in [−Choice] contexts may be a manifestation of the Complexity Principle. Negated predicates are syntactically less simple, less basic, and less canonical than affirmative predicates (Givón 2001: 105; Huddleston and Pullum 2002: 46). Thus, the Complexity Principle predicts that with other things equal, negated matrices should exhibit a higher infinitive-to-gerund complementation ratio than affirmative matrices. This is what we have observed in [−Choice] contexts. It is the disappearance of this effect in [+Choice] contexts that poses a dilemma.

Even more difficult to explain is the phenomenon observed in the complementation of *accustomed* in the Hansard dataset. There, it appears that the infinitive-favoring effect of matrix negation is not only canceled in [+Choice] contexts but reversed, to the degree that even the infinitive-favoring effect of [+Choice] is overridden. We have suggested in Chapter 5 that if this reversal effect turns out to be a real phenomenon, it may be due to the opposite epistemic implications of *afraid* and *accustomed*. However, it must be kept in mind that our parameter estimates for the interaction term—even where statistically significant—are imprecise. Matrix negation is fairly infrequent, so precise estimates of the interaction coefficient are difficult to obtain without huge datasets or purposeful sampling. Accordingly, the confidence intervals for the interaction in Fig. 7.1 imply that the true interaction coefficient may very well be the same for all three adjectives.

As for other syntactic predictors of complement choice, intervening material between the head and the complement was found to favor infinitival variants in the two datasets (Chapters 4 and 5) where the phenomenon was sufficiently abundant to be included in the analysis. This is congruent with the Complexity Principle. An interesting detail was the observed difference in effect magnitude between pre-complement insertions (occurring before the infinitive marker or preposition) and those intervening between the introductory element and the verb i.e. after the speaker has committed to the complement choice. Since the former type delays the speaker's commitment to one or the other option, we think it stands to reason that this kind of insertion should exhibit a stronger cognitive complexity effect. Although the effect differential was not in itself statistically significant (the confidence intervals overlapped), we find the difference in the point estimates to be worth stating for the record.

In closing, we believe that this book provides valuable corroboration of numerous syntactic and semantic factors influencing non-finite alternation. In our day and age, with computational power doubling every two years, these is less reason than before to remain limited to traditional univariate methods which, whenever observational data is involved, run a risk of reporting null or spurious effects due to lurking variables. Univariate methods can be valuable in helping to identify new factors potentially impacting variation in complement choice, and work using such methods should not be discouraged. For instance, the Extraction Principle and the Choice Principle were developed in recent years with the help of such methods, and they played a prominent role in this book. Users of such traditional methods should be aware, however, that a univariate association may disappear or change direction after adjustment for a third variable, as we observed with passivization in Chapter 2 and with DCC-nesting in Chapter 4. Multivariate regression analysis is one of several statistical methods which, when correctly applied, can mitigate such risks. The present book is offered as a concrete illustration of some of its possibilities in research on non-finite complementation.

REFERENCES

Givón, Talmy. 2001. *Syntax: An Introduction*, vol. 1, rev. ed. Amsterdam: John Benjamins.

Huddleston, Rodney D., and Geoffrey K. Pullum. 2002. *The Cambridge Grammar of the English Language*. 3rd printing 2010. Cambridge, UK: Cambridge University Press.

Rohdenburg, Günter. 1996. Cognitive Complexity and Increased Grammatical Explicitness in English. *Cognitive Linguistics* 7 (2): 149–182.

Rudanko, Juhani. 2015. *Linking Form and Meaning: Studies on Selected Control Patterns in Recent English*. London: Palgrave Macmillan.

Vosberg, Uwe. 2003. Cognitive Complexity and the Establishment of -ing Constructions with Retrospective Verbs in Modern English. In *Insights into Late Modern English*, ed. Marina Dossena and Charles Jones, 197–220. Bern: Peter Lang.

INDEX

© The Editor(s) (if applicable) and The Author(s), under exclusive license to Springer Nature Switzerland AG 2021
J. Ruohonen and J. Rudanko, *Infinitival vs Gerundial Complementation with Afraid, Accustomed, and Prone*,
https://doi.org/10.1007/978-3-030-56758-3

159

CPSIA information can be obtained
at www.ICGtesting.com
Printed in the USA
LVHW081231071220
672952LV00036B/459

9 783030 567576